THE Happy Atom STORY 2

READ
A Fantasy Tale
LEARN
Basic Chemistry

BOOK 2

IRENE P. REISINGER
ILLUSTRATIONS BY SARA K. WHITE
GRAPHICS BY IRENE P. REISINGER

Archway Publishing books may be ordered through booksellers or by contacting:

Archway Publishing
1663 Liberty Drive
Bloomington, IN 47403
www.archwaypublishing.com
1 (888) 242-5904

ISBN: 978-1-4808-8316-1 (sc)
ISBN: 978-1-4808-8315-4 (hc)
ISBN: 978-1-4808-8317-8 (e)

Print information available on the last page.

Archway Publishing rev. date: 12/23/2019

The Happy Atom Story

A Four Book Series

Book 1 Guy Learns About the Periodic Table

Guy is introduced to chemistry's unique vocabulary. He learns to interpret the Periodic Table to find information about the atom and the particles in the atom. This begins his Chemistry adventure.

Book 2 Guy's Adventures With the Elements

Using the Periodic Table, Guy draws Bohr models of the atoms. Then he meets the elements in each of the Chemical Families who will be part of his Chemistry experience. In the end Sodium, the scientist of Periodic Table Land, discovers the secret of compound formation.

Book 3 Guy Learns About Compounds and Formulas

Guy learns how Compounds are formed as the elements become Happy Atoms. As Guy watches the compounds forming, he learns the meaning of chemical formulas. Finally he learns to construct chemical formulas using the Valence Method.

Book 4 Guy Learns to Balance Equations

Polyatomic Ions are formed and save the Chemical Families in Periodic Table Land. Guy learns how they form compounds. After that, compounds join to form different kinds of Chemical Reactions. Guy learns how to write Chemical Equations and Balance them. The Book ends with a parade of the principles of chemistry to remind the reader of all the chemistry learned.

DEDICATION

TO

Sister Regina Mercedes, CSJ

My Chemistry Teacher
St. Brendan's High School
Brooklyn, New York

and

John Carlin Ph.D.

My Inorganic Chemistry Professor
Fordham University
New York City

Thank you for helping me understand Chemistry.

TO

MY LOVING PARENTS

Irene and John Murray

Thank you for my happy childhood.

TO

My wonderful husband, **Fred** and my five loving children
Terry, Mary, Kathy, Freddy and John
With All My Love

THE INTRODUCTION

A Fantasy Tale Teaches Basic Chemistry

If you have read Book 1 in *The Happy Atom Story,* series, you know that the principles of chemistry are woven into a fantasy tale for two reasons. First and foremost, the story makes the rather difficult principles of chemistry more understandable. Secondly, the story captures and holds the attention of young readers long enough to have them learn the principle of chemistry being explained. Finally, stories are easier to remember than facts. As the reader remembers the story the chemical principles are recalled along with the story.

The story originated while I was teaching a class of 8th graders how compounds are formed by sharing or exchanging electrons. It was right after lunch and the A/C was not working. Getting their attention seemed a remote possibility. I knew if I didn't do something quickly, they were not going to learn compound formation, and the rest of the unit depended on understanding this. When I gave up the traditional way of explaining compound formation, and told the class a story about sad little atoms that learned to be happy, I had the attention of every student in my class. At the end of the lesson that day, every single student had learned that these sad little atoms became happy when they shared or exchanged electrons. When the atoms became happy they became a beautiful new creation called a compound. This story about compound formation was so effective I continued using it for the next 18 years.

Again if you have read Book 1, you know *The Happy Atom Story* is about a boy named Guy vacationing on the mountain for the summer with his family. He loves the stars, planets and nature all around him and is in awe of his magnificent world. When he wishes on a star, Wish Star appears beside him and tells Guy that chemistry will teach him about a whole new world hidden under the surface of the visible world he loves.

Wish Star begins Guy's introduction to the world of chemistry, but he can't stay all summer. So, he introduces Guy to Professor Terry. This is when Guy's real adventures begin. Professor Terry has a magic Periodic Table in her lab that has an entrance into the fantasy world of Periodic Table Land. Guy's adventures take place in this mystical world. It's here that the elements, the silly electrons, the proper protons and all the little atoms eagerly share their knowledge of the world of chemistry with Guy. The reader learns chemistry along with Guy. When Book 1 ends, Guy knows how to interpret the Periodic Table to help him understand the structure of the atom.

This is Book 2. Because Guy now understands the Periodic Table he is ready to draw the structure of the atom. Once Hydrogen teaches Guy to draw a Bohr model, he then goes about teaching the first twenty elements how to draw their own Bohr models.

Next Guy meets the Chemical Families, and gets to know something about each of the elements in these families. He learns that elements in the same family that appear so very different are really alike in a special way. That's why they are in the same

Chemical Family. Meeting the different Chemical Families and the elements in these families is necessary to understand how compounds are formed in Book 3.

Book 2's final adventure starts with Professor Terry learning from Sodium that all the elements in his family are very sad. This conversation leads Sodium to search all of Periodic Table Land to discover how elements can become Happy Atoms. How atoms become happy is the information needed to understand compound formation, the subject of Book 3. So, Sodium's discovery of how sad little atoms can become Happy Atoms lays the groundwork for understanding Book 3, and brings Book 2 to an end.

Book 3 is about compounds and constructing chemical formulas. Book 4 shows how the chemicals combine to form chemical reactions which are described by equations. Balancing these equations is explained using see saws and the cute little elements who have been part of Guy's adventures from the beginning. Chemistry is not difficult, it's just finding the right medium to convey the principles. This book does just that.

The Happy Atom Story was a success in helping students learn chemistry, and my colleagues never gave up urging me to publish it. Those in the Special Education Department who observed their students understanding chemistry and constructing chemical formula, were insistent that I should publish *The Happy Atom Story*. Barbara Brooks who worked closely with these students said often in her very quiet way, "If you publish your story so many students will be helped." My middle school students loved the story and excitedly added touches to it over the years.

One day by chance, I ran into a student I had taught in middle school. At the time of our meeting .she was then a senior in college. She told me how my story affected her life. "In freshman year, I decided to drop chemistry and give up my dream of becoming a doctor. I went home and thought, why can't my professor make chemistry easy like Mrs. Reisinger." She then said, "I thought about how you used happy atoms to explain chemistry. Through your story I was able to figure out what my 'impossible to understand teacher was trying to say. I passed that course, stayed in the pre-med program. Last week I received my acceptance into medical school. Thank you, Mrs. Reisinger."

That student's story taught me that my Happy Atom Story had a value beyond the middle school classroom. I knew reading my story would help teachers with a different approach to reach more middle school children learn chemistry. It would also help more students succeed in high school chemistry. Her story made me realize that reading my book could help students at any level. *The Happy Atom Story*.provides an overview of basic chemistry, giving students a chance to better understand chemistry when taking higher level classes. From day one they have a point of reference to see where the principles the teacher is explaining fit into the bigger picture of chemistry.

My book, *The Happy Atom Story* can be an asset to teachers who find students having difficulty understanding certain principles of chemistry. It provides a fresh approach to these difficult principles, that students just don't seem to get, explaining them in the traditional way

About the Author

The author was able to write this book because she has a thorough understanding of basic chemistry, the middle school child, and an imagination. Her education plus her life experiences prepared her to write this story. Her background in chemistry gave her the ability to turn the principles of chemistry into a story without distorting their meaning. She received the Gold Medal for highest honors in chemistry at graduation from Fordham University where she earned her undergraduate degree with a double major: Chemistry and Education. After graduation, she passed the New York City exam to become a licensed chemist. She worked as the chemist with a team of doctors doing research on the kidney for two years at New York University Medical Center in the Department of Renal Physiology. During this time she earned her Masters Degree in Science Education from New York University, Washington Square. Emphasis there was on creativity. Next she taught science for a short time in a junior high school in Brooklyn before marrying her college sweetheart who had now become an officer in the US Marine Corps.

The next part of her life contributed to her story-telling ability that was needed to create *The Happy Atom Story.* Her marine husband was gone six months to a year at a time giving her the opportunity to be creative raising her five little Reisingers. At one point she created a kindergarten in her home for Kathy, her third child, and nine other neighborhood five year olds. Then, while her husband worked at NSA and all her children were in school, she taught for five years in the Primary Department of Trinity Lower School in Howard County, Maryland. There she taught second graders how to write three paragraph compositions motivated by pictures. Her science background was utilized by having her enrich the science program for the Primary grades. All this definitely played a part in developing her imagination.

At that time her friends who were chemistry majors along with her in college were teaching in medical schools and universities. She wondered what she was doing involved with young children. Yet she loved what she was doing. Little did she know that she was being groomed to one day successfully teach chemistry to middle school students and eventually author this book. It is amazing how life experiences all blend together to be the perfect preparation for some future achievement as yet unknown. Her involvement with young children was preparing her to one day come up with the Happy Atom Story in that middle school classroom and to write this book.

When her husband retired she began her career teaching middle school. She taught a course whose objective was to prepare middle school students to succeed in high school chemistry. She taught the middle school students basic chemistry for 19 years. It was in that 8th grade classroom that she discovered an extremely effective way to maintain the attention of middle school students, and at the same time have them learn chemistry. This was accomplished by weaving the principles of chemistry into a story. This story was used and refined over a period of 18 years and became the basis of this book, *The Happy Atom Story.* Without her life experiences and her knowledge of Chemistry this could never have happened.

Editing for Science Integrity

The background of the author was reported to assure the reader that the book contains valid science. The author's degree in chemistry provided her with far more knowledge than she needed to teach basic chemistry. For her undergraduate degree she studied Inorganic, Organic and Physical Chemistry. She also had Qualitative and Quantitative Analysis which prepared her to be a chemist. But to assure you that the principles as explained in this book are accurate, she invited two chemists to read and critique her manuscript. It passed their inspection. The backgrounds of these chemists are described next.

Joyce M. Donohue, Ph.D.—taught chemistry on every level: high school, community college, university and graduate school. Presently, she's working for the EPA doing long term studies on water pollutants. In addition she is teaching chemistry in a local community college. Having taught high school Chemistry, Joyce's final comment was, "High school teachers will really appreciate the way your book prepares their students."

Mr. Robert L. Zimmerman, Jr., M.S.—is a chemist who was the Director of the Research Lab for Customs and Border Protection, a part of Homeland Security, until he retired. His new career, working for the American Association for Laboratory Accreditation to assess laboratories throughout the country to the *International Standard for Chemistry Labs*. After reading *The Happy Atom Story,* he liked especially the study suggestions the book provides.

I appreciated these professionals taking the time to read my manuscript assuring you that you are reading solid basic chemistry. It is important that a reader knows from the outset that the science they are reading is valid. This has been verified.

Success Stories with High School Students

Kati Yost was another success story. Her mother had heard that chemistry was a really challenging subject. She heard I was writing a book about chemistry, and asked me to help. Kati came and read my manuscript of *The Happy Atom Story* twice a week during the summer before she was to take high school chemistry. Helping Kati provided me an opportunity to settle the question: "Can a student read this book without teacher intervention and still understand it?" I knew that using the story involving happy atoms helped students when I was there explaining it in the classroom. I now needed to know if reading my book, *The Happy Atom Story* would be as useful a tool as the story was in the classroom. Kati read the book at my house each week; and after she read it, I tested her on the chemical principles she read. She proved that she thoroughly understood what I wrote without a teacher explaining it. Kati's contribution proved that the book can be read and understood without a teacher. Kati went on to achieve an A in high school Chemistry. Her mother Jane Yost credited *The Happy Atom Story* for helping Kati succeed.

My granddaughter, Meredith Reisinger came to visit one weekend when she was taking high school chemistry. She was stressing about the thought of the test on compound formation that she was facing on the Monday right after her visit. I took her aside and explained a short version of the Happy Atom method of constructing chemical formulas. She aced the test.

Groups That Should Consider Buying *The Happy Atom Story*

Parents

My neighbor is a parent who wants to see her child succeed. When I told her about the Happy Atom story she said, "Write that story in a hurry. My daughter is taking AP Chemistry in the fall. My husband and I do not remember enough chemistry to help." So she pointed out that this kind of a book would be eagerly received by parents who wish to help their child in such a challenging subject.

My granddaughter Kristine's husband, Phil has a Ph.D. in chemistry and works as a research chemist for 3M. He is interested in getting hold of my book and using it to introduce chemistry to his three little children. He plans to read parts of it to them before they learn to read, letting them enjoy the pictures as he tells the story.

The family across the street has a child who is an avid reader. Both his parents are physicians They want my book to encourage their son to extend his interest in science to include chemistry.

Home School Groups

There are several home school communities in my area who are interested in my book. They would like to use it at a young age to help their students become grounded in the basics of chemistry before they take high school chemistry. It can also help teachers get another perspective on chemistry, especially if they were not chemistry majors in college. I've spoken to many other parents who home school their children, and they see a great need for this kind of a book in their home school communities.

Students at Any Level

Students at any level—middle school, high school, community college or even the university.would do well to read *The Happy Atom Story* during the summer before taking chemistry. The book provides an understandable framework on which to hang the many more technical concepts addressed in higher level classrooms. All students will benefit from looking at chemistry from a new perspective.

Special Education Teachers

There are requirements presently for special education teachers to demonstrate how they are differentiating their lessons to accommodate different learning styles. *The Happy Atom Story* definitely does this in the field of chemistry. The reader will appreciate how this book makes theoretical subject matter visual and more understandable. Also stories help students understand concepts with explanations on their level, and stories help students remember the chemical principles longer.

Teachers with Chemistry in their Curriculum

Many 6th Grade teachers have a unit on chemistry in their curriculum. *The Happy Atom Story* brings chemistry down to a level their students can understand. For higher level chemistry teachers it affords a fresh perspective on how to explain principles that students are having trouble understanding.

College Students with Chemistry as a Requirement

Many college students, who are not chemistry majors, are required to take chemistry for their degree. This is often a real challenge. By reading the book they will be establishing a base to build on before they take the higher level chemistry course.

Students studying for a degree in Science Education would especially profit from reading *The Happy Atom Story.* This book would be a good resource if they some day have to teach chemistry as part of a science class.

The General Public

Educators hope that the general public will increase their knowledge of science so they can have a better understanding of how science affects their lives. Research has shown that chemistry is the least understood of all the sciences. Reading *The Happy Atom Story* will give the reader an overview of what is basic to chemistry. I've spoken to many obviously intelligent people at different events and most of them verbalized that they had a hard time understanding what chemistry was all about when they took the class in high school. Others said, "I flat out failed that subject." One said, "I took the course and got a good grade, but I never did figure out what it was all about. I just memorized everything." I've talked about the book with my doctors, and they agreed that chemistry was a difficult subject, and a book like mine is seriously needed. One podiatrist said, "I'd read it just for the heck of it."

Foreign Countries

The internet indicates foreign countries are looking for science books written in lay man's language. *The Happy Atom Story* fills this need in the area of basic chemistry.

BOOK 2
Part 1
THE BOHR MODELS

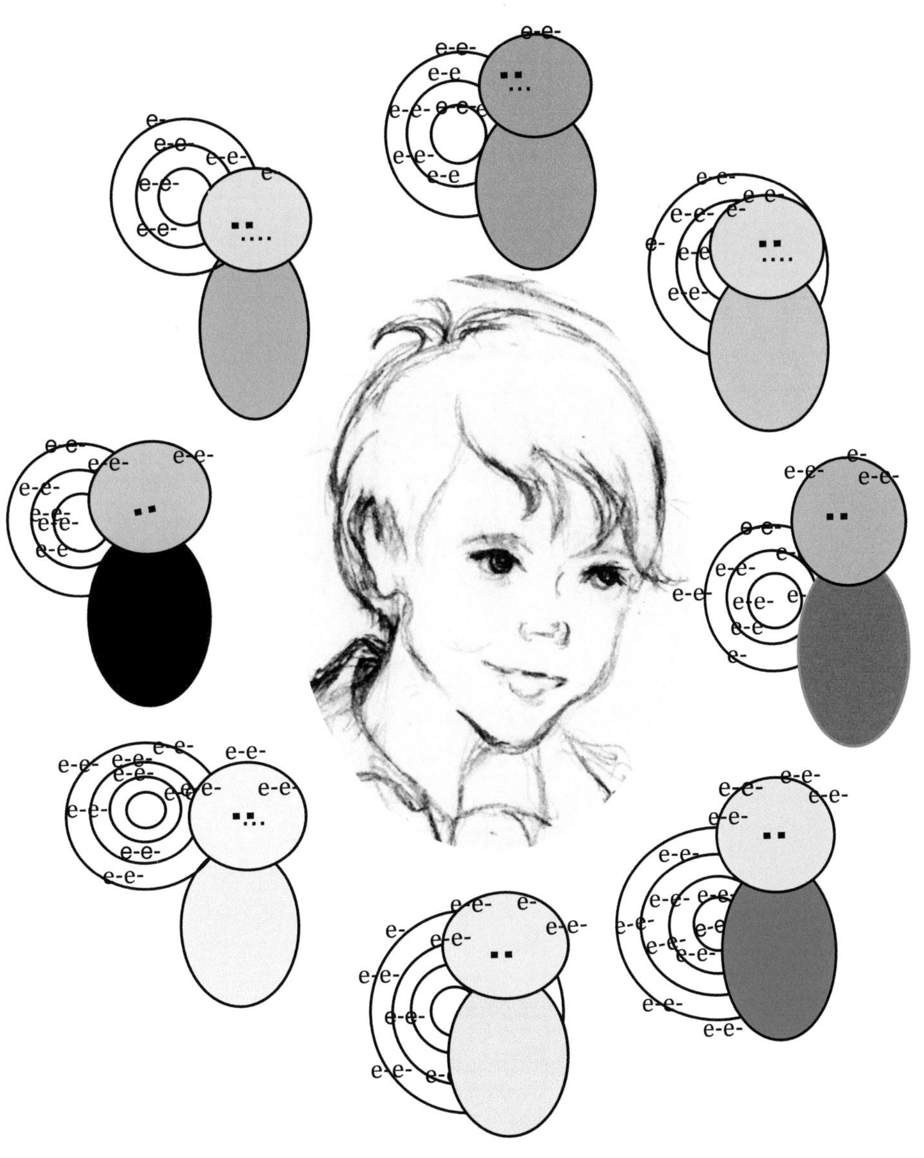

HELPFUL CHARTS

PERIODS

GROUPS →

THE PERIODIC TABLE OF THE ELEMENTS

Periods	1A	2A	3B	4B	5B	6B	7B	8B	8B	8B	1B	2B	3A	4A	5A	6A	7A	8A
1.	1 **H** 1.00																	2 **He** 4.00
2.	3 **Li** 6.94	4 **Be** 4.01											5 **B** 10.8	6 **C** 12.0	7 **N** 14.0	8 **O** 16.9	9 **F** 18.9	10 **Ne** 20.1
3.	11 **Na** 22.9	12 **Mg** 24.3											13 **Al** 26.9	14 **Si** 28.0	15 **P** 30.9	16 **S** 32.0	17 **Cl** 35.5	18 **Ar** 39.9
4.	19 **K** 39.1	20 Ca 40.0	21 **Sc** 44.0	22 **Ti** 47.9	23 **V** 50.9	24 **Cr** 51.9	25 **Mn** 54.9	26 **Fe** **56**	27 **Co** 58.9	28 **Ni** 58.6	29 **Cu** 63.5	30 **Zn** 65.3	31 **Ga** 69.2	32 **Ge** 72.6	33 **As** 74.9	34 **Se** 78.9	35 **Br** 79.9	36 **Kr** 83.7
5.	37 **Rb** 85.5	38 **Sr** 87.6	39 **Y** 88.9	40 **Zr** 91.2	41 **Nb** 92.9	42 **Mo** 95.9	43 **Tc** {98}	44 **Ru** 101	45 **Rh** 103	46 **Pd** 106	47 **Ag** 107	48 **Cd** 112	49 **In** 114	50 **Sn** 119	51 **Sb** 122	52 **Te** 126	53 **I** 126	54 **Xe** 131
6.	55 **Cs** 132	56 **Ba** 137.	57 - 71	72 **Hf** 178	73 **Ta** 180	74 **W** 183	75 **Re** 186	76 **Os** 190	77 **Ir** 192	78 **Pt** 195	79 **Au** 196	80 **Hg** 200	81 **Tl** 204	82 **Pb** 207	83 **Bi** 208	84 **Po** 209	85 **At** 210	86 **Rn** 222
7.	87 **Fr** 223	88 **Ra** 226	89-103	104 Rf 267	105 **Db** 268	106 **Sg** 271	107 **Bh** 272	108 **Hs** 270	109 **Mt** 278	110 **Gs** 281	111 **Rg** 280	112 **Cn** 285	113 Nh 284	114 **Fl** 289	115 **Mc** 288	116 **Lv** 203	117 Ts 294	118 Og 294

57 **La** 139	58 **Ce** 140	59 **Pr** 141	60 **Nd** 144	61 **Pm** 145	62 **Sm** 150.	63 **Eu** 151	64 **Gd** 157	65 **Tb** 159	66 **Dy** 163	67 **Ho** 165	68 **Er** 167	69 **Tm** 169	70 **Yb** 173	71 **Lu** 175
89 **Ac** 227	90 **Th** 232	91 **Pa** 231	92 **U** 238.	93 **Np** 237	94 **Pu** 234	95 **Am** 243	96 **Cm** 247	97 **Bk** 247	98 **Cf** 251	99 **Es** 252	100 **Fm** 257	101 **Md** 258	102 **No** 259	103 **Lr** 262

Professor·Terry's·Element/Symbol·Chart¶

Atomic·Number/·Name·of·Element/·Symbol

1	Hydrogen	H	8	Oxygen	O	15	Phosphorus	P
2	Helium	He	9	Fluorine	F	16	Sulfur	S
3	Lithium	Li	10	Neon	Ne	17	Chlorine	Cl
4	Beryllium	Be	11	Sodium	Na	18	Argon	Ar
5	Boron	B	12	Magnesium	Mg	19	Potassium	K
6	Carbon	C	13	Aluminum	Al	20	Calcium	Ca
7	Nitrogen	N	14	Silicon	Si			

BOOK 2
Part 1
Bohr Models

Guy Learns to Draw Bohr Models

Chapter 1

The sun peeked over the mountain tops and sent its early morning rays through the cabin's bedroom window to playfully dance around the sleeping boy's face. Guy snuggled deep into the folds of his down comforter, resisting the sun's message to get up. Slowly, he opened his eyes and lay there for a while in his soft comfortable bed. He smiled enjoying happy memories of Periodic Table Land: sliding through the mirrored tunnel with a thousand sparkling lights; silly electrons teaching him about their negative charges; protons rolling up play dough to teach him about their buddies the neutrons; his trips with Sodium flying around in the magic bubble to visit all those elements. As he became more awake, he thought of the exciting day ahead. Today he would see what the elements' atoms were like. He wondered what his Bohr model adventure would involve. He was brought back to reality as the aroma of his mother's freshly baked cinnamon buns wafted into his room. Hearing the bacon sizzling on the stove motivated him to begin the day. He jumped out of bed, pulled on his clothes and joined his parents who were already drinking their morning coffee at the kitchen table.

Guy's father greeted him with an approving smile saying, "Professor Terry called and told me about your great progress learning chemistry. She said you know all about the Periodic Table, and you're ready to draw Bohr models of the elements. We're so proud of you ,Guy. What makes us even happier is that you are having such a good time in Professor Terry's fantasy world in addition to learning chemistry." The family sat around enjoying the delicious breakfast food and sharing their plans for the day until it was time for Guy to leave.

As Guy left his family's summer cabin, he felt the coolness of the early morning mountain air that was so refreshing. The dew on the grass and the morning glories just opening their beautiful blue trumpets reminded Guy how much he loved all of nature on his mountain. He realized how much more he needed to learn about the world of atoms hidden deep below the surface of this world. The morning glories called after him, "Run, Guy, run. We want you to learn about the atoms that make us who we are." They reminded Guy that when he learned about their atoms, he would be learning about the atoms that make up the entire universe. For the rest of the way down the mountain, Guy thought of nothing but the models of the atoms that he was about to draw—the atoms which would give him a peek into the universe.

When he arrived early at the university lab, he was ready to run in and start drawing Bohr models right away. Well, that didn't happen. When he arrived, Professor Terry was involved in an experiment. Guy usually loved watching her work, but today he was impatient. Noticing how unhappy he was feeling, he remembered something his grandmother had told him, "Patience is a virtue, Guy. You'll be a happier person if you learn to be patient. Life gives you many opportunities to practice patience. Be happy." This was definitely one of those moments she talked about as he noticed again how unhappy he was waiting. Guy reflected, "Being impatient isn't getting me to draw Bohr models any sooner. So I may as well be happy while I wait." He decided to enjoy watching Professor Terry experimenting.

Professor Terry invited Guy to come closer. She gave him safety glasses to wear like the ones she had on. She told Guy she was performing the last of a series of tests on a sample sent to her for analysis. Her job was to identify its contents. She said, "The results of my analysis are due today. Remember how you learned about pure substances and mixtures. My clients want to be sure they are buying pure substances not mixtures. So I have to complete what I'm doing before I can give you my attention. I will be finished in a little while. Then you can begin your Bohr model adventure." Guy continued to watch the experiment in progress, fascinated to be in a lab with a real experiment being performed. Professor Terry finally finished.

She said, "I know you're really excited about getting to draw Bohr models. You will get to see the structure of the atoms of all those elements you met, and some you haven't met as yet. Drawing Bohr models also will give you practice in one of the essential habits chemists must develop—being orderly and systematic. It helps chemists avoid mistakes. If they work in an orderly way, they can find where they went wrong without starting all over. It saves so much time. It can also help you in your private life to be orderly and systematic. If you put things back where they belong, you will cut down on how much time you spend searching for lost items. Think about this. You will begin today becoming systematic as you record the work you do while drawing Bohr models. You can start by creating a three column chart on your pad with these headings:

ELEMENT BOX CALCULATIONS BOHR MODEL

"When I'm finished making sure you're prepared, we'll be leaving to see Hydrogen—Atomic Number 1. He will help you draw your first Bohr model. First you need to locate your Periodic Table and Element/Symbol Chart. You will need to have these two charts handy while you are drawing Bohr models.

. Guy searched through his important papers' file and found the Periodic Table and his Element/Symbol Chart. which lists the first 20 elements with their symbols and Atomic Numbers. He had memorized these symbols when he first met Professor Terry. He smiled as he remembered this was how his whole adventure began. He looked once more at the chart that helped him start to learn chemistry. He thought, "Now, I will use the Atomic Numbers on this chart to find the element boxes on the Periodic Table. He placed this chart along with his Periodic Table in a new folder, and slid it into his backpack. He was ready. Guy said, "I've got everything needed. I'm ready to go"

Professor Terry said, "One last thing, I'd like to do a short review with you. Do you remember the formula to calculate neutrons?"

Guy said, "How could I forget it! After that play the atoms put on for me. I'll always remember the formula for neutrons, $n^0 = A - Z$."

"**First** you find the number at the top of the element box. This is the **Z** number that you need for the formula. **Second,** you round off the *amu* at the bottom of the element box. This gives you the **A** number you need in the formula. **Third,** you do the math. Now that you know the number values for A and Z, you subtract them. This will give you the number of neutrons in the nucleus of the atom. Last, you write $n^0 =$ and you write the number of neutrons that you got when you used the formula $n^0 = A - Z$. You subtracted the number value of Z from A and got the number of neutrons to put in the atom's nucleus. Then you are finished."

"When you know how to do it, it's easy. I remember when I didn't know how to calculate the number of neutrons in an atom. I wondered if I'd ever learn. I was especially doubtful when I heard the neutrons were not listed in the element box. I guess I am ready to go over to Periodic Table Land now. I'm ready to have Hydrogen teach me how to draw Bohr models."

Professor Terry was pleased with all Guy's answers and agreed, "That's it Guy! You are ready to draw Bohr models."

Hydrogen Teaches Guy to Draw His First Bohr Model

They went to the back of the lab, pushed against the Periodic Table. Bells, whistles and sirens screamed—adventure ahead. Professor Terry said, "Here we go!" As she held Guy's hand they found themselves sliding down the tunnel of mirrors reflecting a thousand sparkling lights. Magical sprinkle dust enveloped them like a silky blanket. They felt comforted by soft mysterious music. Then, in an instant, they were again, in that fantastical world, Periodic Table Land. Guy was always thrilled when he was greeted by those colorful blinking lights surrounding the street signs. These were the Groups on the Periodic Table now turned into street signs. There were 8 streets in Periodic Table Land and a wide B Avenue in the middle of the town. The streets were the A Groups on the Periodic Table. What were formally the boxes on the Periodic Table are now the most adorable houses displaying so many different shapes and sizes. Guy noticed the elements running around. He looked up and saw they had arrived. Professor Terry and Guy were at Hydrogen's door. He was expecting them.

Hydrogen threw open the door and welcomed them into his comfortable home. They sat down and chatted a while. Then Hydrogen eager to begin said, "OK Guy, first thing you need to do is take out your Element/Symbol Chart. You'll need the Atomic Numbers to find the elements on the Periodic Table. I hear you know how to use the Periodic Table to find the particles in the atom."

Guy opened his folder of important papers, took out his Periodic Table and placed it where it would be handy as he drew the Bohr model. Guy said, "If I know the atomic number of an element, I'll be able to find it on the Periodic Table." Then he listened carefully to the instructions as Hydrogen directed him to draw his first Bohr model.

"In your column marked ELEMENT BOX copy my Hydrogen box from column 1A on the Periodic Table, and then label it as I direct you. Next to the Atomic Number put its symbol Z and the particles that equal the Atomic Number, the electron e^- and the proton p^+. Then Guy, label the *amu*, that long decimal number at the bottom of the element box, 1.00794. Now round off the amu to get the atomic mass, A. 1.0 = 1.A = 1

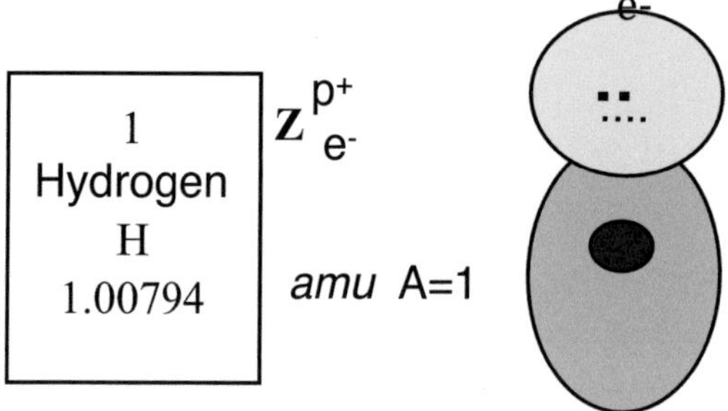

Hydrogen smiled as he looked approvingly at Guy saying, "You both copied and labeled the element box correctly." Then he continued his directions, "Next, put the formula to solve for neutrons in the CALCULATIONS column." Guy wrote:

CALCULATIONs

$$\mathbf{n^0 = A - Z}$$

Then Hydrogen continued. "The next step is called *Substitution.* All you need to do is to *substitute numbers for the letters in the formula."*

Guy thought out loud as he looked at Hydrogen's element box. "The atomic mass number, A is 1, so I put a 1 under the A. The Atomic Number, Z is 1. So I put a 1 under the Z. That's substitution. Next I do the math. 1-1=0. The answer is 0. There are 0 neutrons in the nucleus of the Hydrogen atom. So $\mathbf{n^0}$ = 0 Here's the calculations.

Formula	$\mathbf{n^0 = A - Z}$
Substitution	$\mathbf{n^0 = 1 - 1}$
Answer	$\mathbf{n^0} = 0$

Hydrogen looking over Guy's work said, "Now you found out why I am the lightest element in the universe. Remember the mass of the atom is equal to the neutrons plus the protons. I have no neutrons and all the other elements have neutrons. They also have more protons than I do because all their Atomic Numbers are larger than mine which is 1. So if the mass is equal to protons plus neutrons, the mass.of all the other elements is greater than mine. So those are two reasons that I'm the lightest element in the universe. Now, let's draw your first Bohr model."

Hydrogen continued, "In the **BOHR MODEL** column draw a small thick circle for the nucleus." Hydrogen explained, "We make the circle for the nucleus thick so that it will not be mixed up with the energy levels.

Then, he asked, "What particles are in the nucleus of every atom?"

Guy responded, "Protons and neutrons. You have no neutrons, so in your nucleus there is only a proton. Since the number of protons is equal to your Atomic Number, that means you have 1 proton in your nucleus." He wrote in the circle 1 p^+

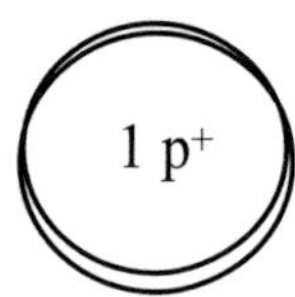

Hydrogen continued, "Draw a circle around the nucleus for the energy level. Then, check to see how many electrons you have to put in my energy level."

Guy responded, "The number of electrons is equal to your Atomic Number. So, I will put 1 electron$^-$ in that energy level."

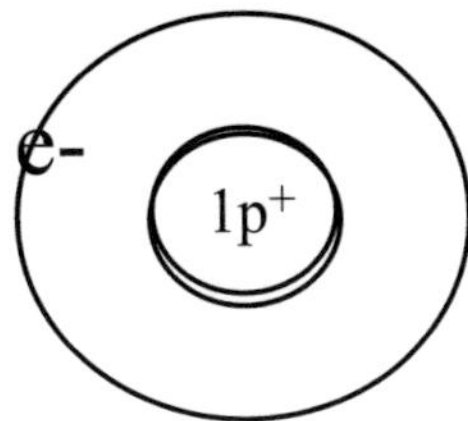

Hydrogen said, "The moment you put that 1 electron in my energy level, you are finished drawing my Bohr model because my Atomic Number is 1. I have only 1 electron. You've done it! You have drawn your first Bohr model." Hydrogen checked the finished product and was pleased with the results.

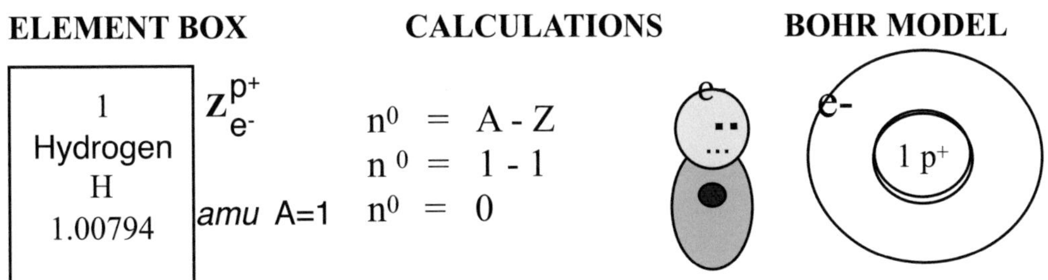

"**Hydrogen**, notice that you can look at the Bohr model and it tells you the exact information as your body markings and your element box on the Periodic Table"said Professor Terry and she continued. "Your 1e hair electron says you have one electron in your Outside Energy Level. Check your model. You do. The dot on your body says you have one energy level. Again you do. Your Atomic Number in your Element Box says you have 1 electron and 1 proton and you do. Amazing what your Bohr model can tell you." Hydrogen was impressed. Then Professor Terry and Guy got ready to leave.

"Thank you Hydrogen for being such a good instructor." She asked to borrow his family's magic bubble to get around saying, "We will need a way to fly around Periodic Table Land as Guy draws Bohr models for the rest of the elements, Atomic Numbers 2 through 20."

Hydrogen said, "You can use the magic bubble, but don't leave yet. Before you go, I want to tell you a little about my element. Since you finished drawing my Bohr model, you should be ready to learn all about me. Let's sit down and visit a little while."

Professor Terry and Guy saw a comfortable bench right outside Hydrogen's house. They sat down and Hydrogen began to tell them all about his element and some compounds he enjoyed making.

HYDROGEN

"You already know I'm the lightest gas in the universe, but I didn't tell you that because I am so light most of my atoms fly off into outer space. That's why you don't see pure atoms of Hydrogen gas on Earth. However, I'm a big part of your sun and all the other stars in the universe. Chemists have to make Hydrogen in the lab when they need me." Continuing he said, "They tried to use me in air ships because I'm the lightest element; but since Hydrogen is very explosive, it was a bad idea. The Hindenburg, an early air ship, called a blimp, filled with Hydrogen gas exploded and crashed killing 35 people, including those on board and others on the ground. Now they use Helium in air ships instead of me. I was disappointed, but I understood. I have many other uses. I am found in the compound water which covers 3/4 of the earth's surface, and I'm also an important part of acids. I make lots of compounds with Carbon. Some are called *hydro*carbons because they're made only of the elements Hydrogen and Carbon. Here are just a few familiar hydrocarbons: methane which is swamp gas, butane used in lighters, and propane the gas you use to grill hot dogs."

Guy and Professor Terry thanked Hydrogen for his help as they climbed into the magic bubble. She checked to be sure Guy had his important papers. Soon they were lifted by those beautiful light green balloons, and the magic bubble started Guy on his Bohr model adventure. The plan was to draw the Bohr models in order of the elements' Atomic Numbers. So Helium, Atomic Number 2 would be next.

Professor Terry said, "One more thing Guy. When scientists draw the nucleus of the Hydrogen atom, they leave out the 1 and just write p^+. Why? Because when they write just the p^+., it means there is one proton in the nucleus or they wouldn't write anything. So, they just write the p^+, and they don't have to write the one. Do you understand?"

Guy responded, "I understand. Writing the one is just not necessary."

GUY'S FIRST BOHR MODEL!

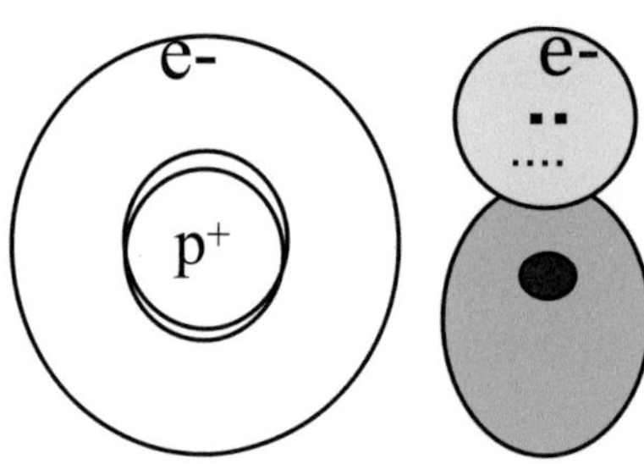

Guy's Adventures Drawing Bohr Model
Chapter 2

The magic bubble floated effortlessly across Periodic Table Land carrying Professor Terry and Guy to 8th Street, the home of Helium, Atomic Number 2. Guy was committed to drawing the Bohr models in strict Atomic Number order. He had just drawn Hydrogen's Bohr Model, Atomic Number 1. So he needed to draw the Bohr Model for element Atomic Number 2 next. That was Helium. This meant he had to travel all the way from 1st Street to the last street in the land. As they flew over the whole country, Guy's eyes searched below for some familiar houses. Silicon's sand castle and Carbon's log cabin were his favorites. Many homes were surrounded by colorful gardens, but none as beautiful as Nitrogen's because he knew how to fertilize the soil in his garden.

They enjoyed looking at all the interesting houses in Periodic Table Land throughout the trip. It seemed like no time when they were landing on Helium's property decorated with his colorful balloons. Helium said, "I'm blowing up a dozen balloons for the element, Fluorine. He ordered them for a party this evening. What can I do for you?"

Guy said, "I'd like to draw your Bohr model. Let me show you the one I drew for Hydrogen." Guy showed Helium Hydrogen's Bohr model and explained, "Your model will be something like this but not exactly. Each element's atom is uniquely different from the atom of every other element. Then Guy showed him the special Steps Chart that he had made. "I've made a chart to show you the steps I follow when drawing Bohr models. As I draw your Bohr model, you can follow along on this chart as you watch\ what I'm doing." Guy began and Helium followed along.

Steps Chart
How to Draw a Bohr Model

1. **Draw three columns**, and label each column.
   ELEMENT BOX CALCULATIONS BOHR MODEL
2. **Copy** element box from the Periodic Table into the ELEMENT BOX column.
3. **Label** the box correctly. Put the symbol Z next to the Atomic Number.
   Next to the Z put (p+) and (e-) because they equal the Atomic Number.
   Next round the *amu*. This becomes the A number, the mass of the atom.
   Write the A number next to the *amu* right outside the Element Box.
4. **Calculate** the number of neutrons (n^0) to put in the nucleus.
   Write the formula, $n^0 = A - Z$ in the CALCULATIONS column; substitute the values for A and Z; do the subtraction. The answer is the number of neutrons (n^0) to put in the nucleus.
5. **Draw** the atom in the Bohr Model column. Put the protons and neutrons in the nucleus, and the electrons in the energy levels.

Guy Gets Busy Drawing Helium's Bohr Model

1. Label the Columns:

ELEMENT BOX **CALCULATIONS** **BOHR MODEL**

2. Copy and Label the Element Box from the Periodic Table.

2 Helium He 4.0078	Z p^+ e^- A=4

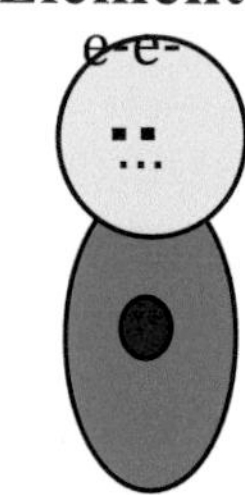

3. Do the Calculations

$n^0 = A - Z$

$n^0 = 4 - 2$

$n^0 = 2$

4. Draw Helium's Atom

1st Draw a thick circle for the nucleus.

2nd Helium is in **Period1. So, Draw** 1 Energy Level around the nucleus.

Helium's Atom

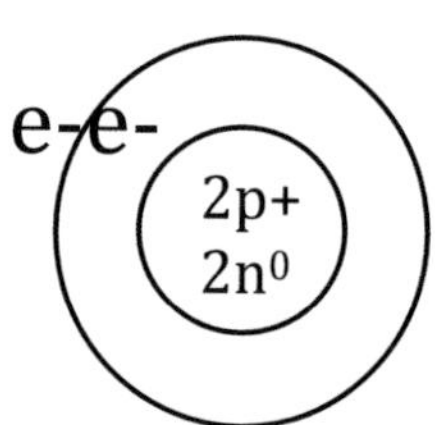

3rd Put the protons and neutrons in the nucleus.

p+ = Atomic # Helium's is 2 2p+

n^0 neutrons (See Calculations). $2n^0$

4th Put electrons in Energy Level.

# of Electrons equal the Atomic Number.

Helium's Atomic Number is 2.

Put 2 electrons in the energy level.

Guy said, "I've finished drawing your Bohr model. Come look at it.

HELIUM'S BOHR MODEL

ELEMENT BOX **CALCULATIONS** **BOHR MODEL**

2 Helium He 4.0078	Z p^+ e^- A=4

$n^0 = A - Z$

$n^0 = 4 - 2$

$n^0 = 2$

e-e- 2p+ $2n^0$

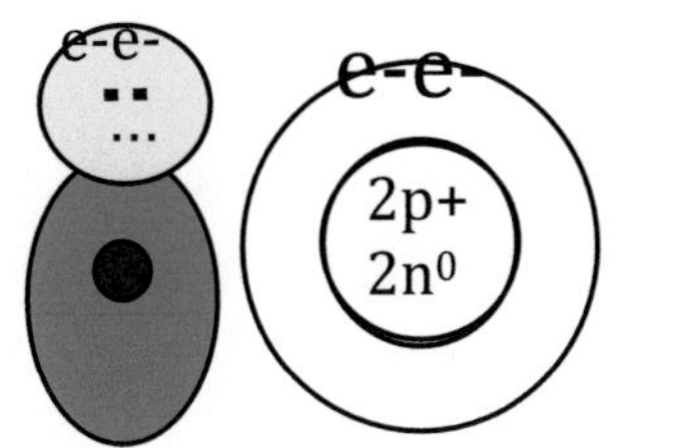

2
Helium
He
4.0078

"**Helium** looked in the mirror and noticed how his Bohr model told the same information as the markings on his body. The one dot on my body says I have one Energy Level, and I do. My 2e hair says I have 2 electrons in my Outside Energy Level, and I do. These same electrons say I am in Group 2A. Amazing, isn't it? It helps me understand my element box too. My Atomic Number is 2, and I do have 2 protons and 2 electrons in my atom. My Bohr model tells a lot."

Helium said, "I love how much my atom tells about me. Now that you've finished drawing and explaining my Bohr model, I would like to have you and Professor Terry sit down and visit with me. I'd like to tell you about myself."

Professor Terry was happy to have Guy learn something about Helium. So they went into Helium's house and sat down and listened.

HELIUM

Helium began sharing his history. "I go way back to ancient times. The Greek word for the sun is helios. I was named after the sun because I am one of the two major gases found in the sun. You know me best because I'm the gas that inflates your Happy Birthday balloons. They started using me in zeppelins (air ships) instead of Hydrogen because I'm the second lightest gas in the universe. But a more important reason why they chose me is because I'm not flammable. I don't burn or explode. I'm used in cryogenics. That's the science of cooling to very low temperatures. I'm used to cool superconducting magnets like the MRI. I have so many uses."

Helium thanked Guy for showing him how to draw his Bohr model. He said, "Now I can look at my box from the Periodic Table, and it is much more meaningful."

Helium was so excited to have learned how to draw his own Bohr model. He couldn't stop looking at it! Guy said, "Now when I look at the sun I'll know that Helium's up there. I even know what Helium's atom looks like. Guy and Professor Terry said, "Good bye." They were off to see Lithium, element Atomic Number 3.

Helium was delighted to see the Bohr model of his atom. He couldn't stop looking at it. He decided to frame it and hang it on his living room wall.

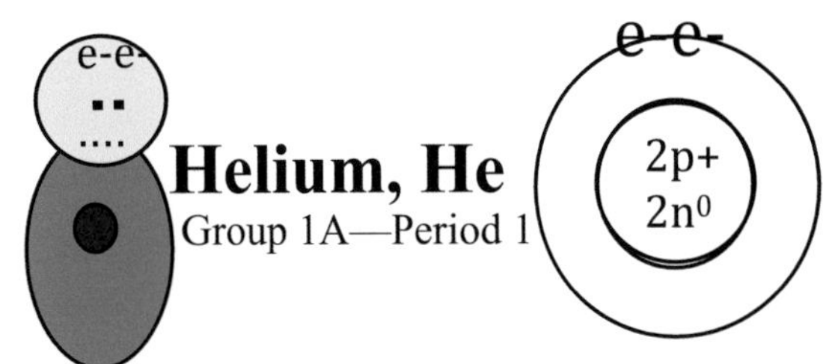

LITHIUM

They jumped into the magic bubble and flew back to 1st Street on the other side of Periodic Table Land to visit **Lithium**, Atomic Number 3. On the way Professor Terry came up with an idea. She said, "I think it's a good idea to keep track of the elements whose Bohr models we've completed." She took out a condensed version of the Periodic Table Land map saying, "This map has only the elements that we will be visiting. I've colored in the elements we have already visited—Hydrogen and Helium. This makes it clear how much we have accomplished. The boxes not colored let us know how many more elements we have yet to visit. Lithium is next. After our visit, we will color in Lithium's box."

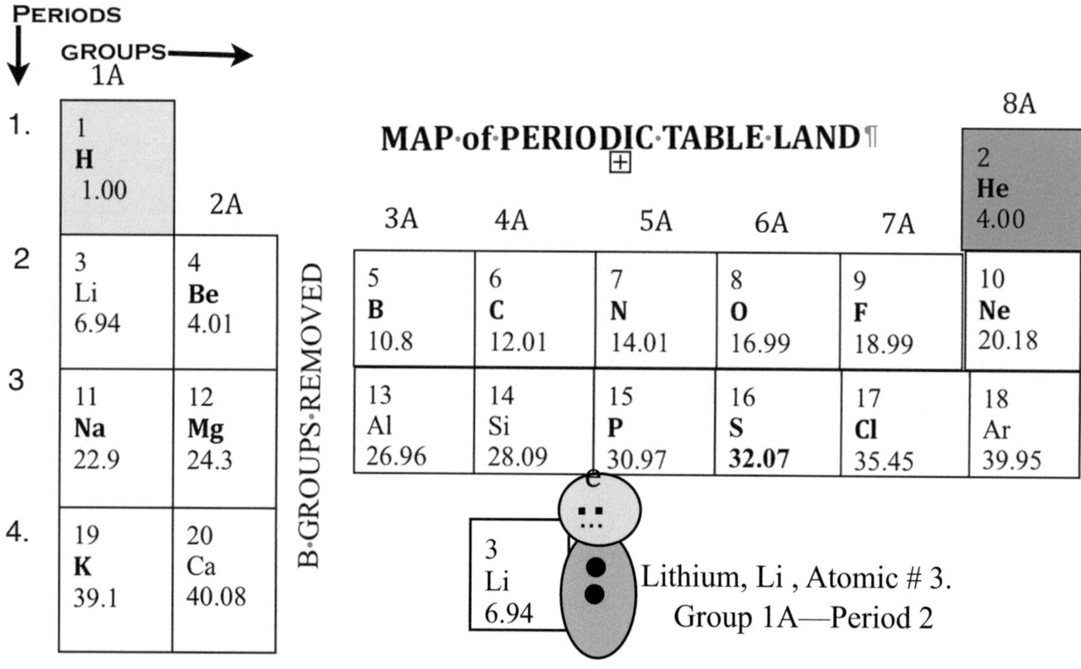

The trip went by fast as they spent so much time checking out the map. They landed in the cleared area behind the trees on 1st Street and made their way down to Lithium's home. #3, 1st Street. His address is #3 because his Atomic Number is 3.

When they got there, they knocked at the door and waited. Soon the door opened. Guy explained that he had come to draw Lithium's Bohr model. Lithium was very interested in learning what his atom looked like.

Guy said, "The Bohr model of your atom will display both the information in your element box as well as the marks on your body.You'll love checking it out."

Guy explained he had brought along the chart that he had made to show the steps he would follow drawing Lithium's Bohr model. Then he hung the chart where **Lithium** could see it as well as what Guy was drawing. Lithium sat next to the chart ready for Guy to begin drawing his Bohr model. Guy was ready and he began. Lithium looked on eager to learn how to draw it himself.

Guy Gets Busy Drawing Lithium's Bohr Model

1. Label the Columns:

ELEMENT BOX **CALCULATIONS** **BOHR MODEL**

2. Copy and Label the Element Box on PeriodicTable.

3 Li Lithium 6.94	$Z^{p^+}_{e^-}$ A = 7

3. Do the Calculations

$n^0 = A - Z$

$n^0 = 7 - 3$

$n^0 = 4$

4. Draw Lithium's Atom

1st Draw a thick circle for the nucleus.

Lithium's Atom

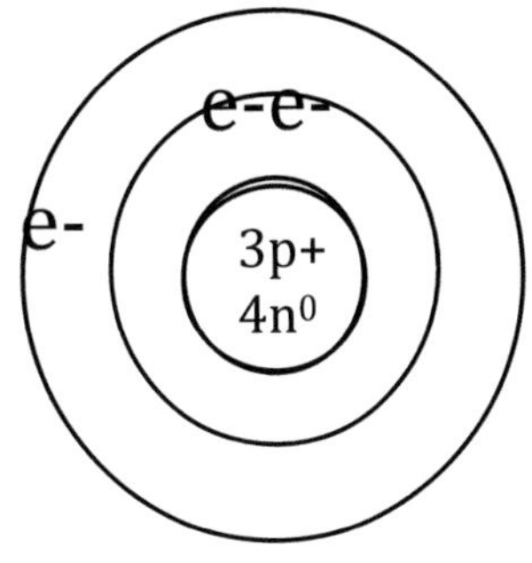

2nd Lithium is in **Period 2,** So draw 2 Energy Levels around the nucleus.

3rd Put the protons and neutrons in the nucleus.

p+ = Atomic # Lithium's is 3 3p+
n^0neutrons (See Calculations). $4n^0$

4th **Put electrons in Energy Levels.**
# of Electrons equals the Atomic Number.
Lithium's Atomic Number is 3.
Put 3 electrons in the energy levels.

Guy said, "I've finished drawing your Bohr model.
Come look at it."

LITHIUM'S BOHR MODEL

ELEMENT BOX **CALCULATIONS** **BOHR MODEL**

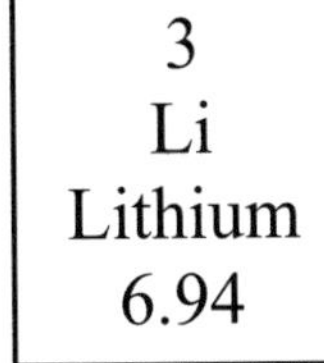

$n^0 = A - Z$

$n^0 = 7 - 3$

$n^0 = 4$

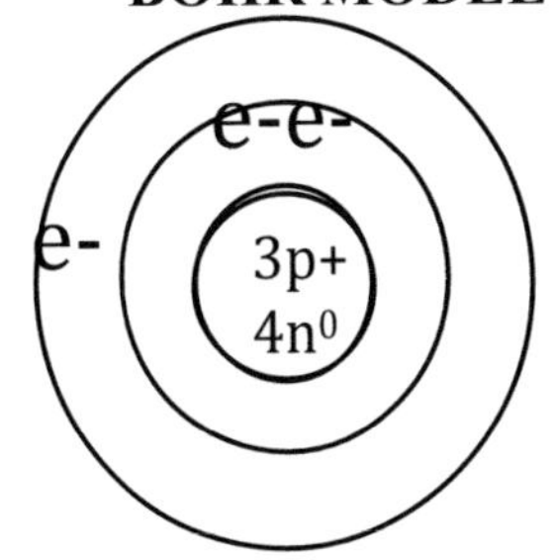

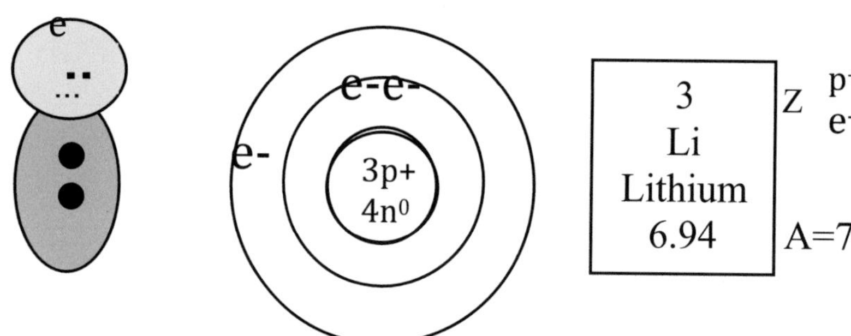

Lithium looked at his Bohr model and said, "Now my element box means so much more to me. I see that my Atomic Number 3 represents the 3 protons I have in my nucleus, and the 3 electrons I have in my energy levels. This picture of my atom makes me understand what being in Period 2 on the Periodic Table means. I do have 2 energy levels in my atom around my nucleus. I can also see why I need 2 energy levels. I have 3 electrons. Since the energy level next to the nucleus can only hold 2 electrons, I need an energy level for my 3rd electron. My body markings make sense now. My 1e hair represents that 1 electron I see in the Outside Energy Level of my atom. The 2 dots on my body mean I have 2 energy levels, and I do. It all makes sense to me now. Bohr models are wonderful.

LITHIUM

Lithium said, "I would like to have you sit down and visit with me. I would like to tell you some more information about me. When I talk of *my* uses, most of the time I am referring to uses of my compounds. Lithium began, "My most important use is in the field of medicine. I'm used in mood stabilizing drugs that help people who are bipolar. Doctors prescribe drugs with Lithium in them to help their patients have more normal lives. I'm also used to make heat-resistant glass and ceramics. I'm also used as an additive in the production of iron, and steel. My use in lithium batteries is widespread. Guy, I bet you know how many toys depend on my batteries to work. You can also think of me the next time you replace a battery in your watch. "

When they left Lithium couldn't stop looking at his wonderful Bohr model.

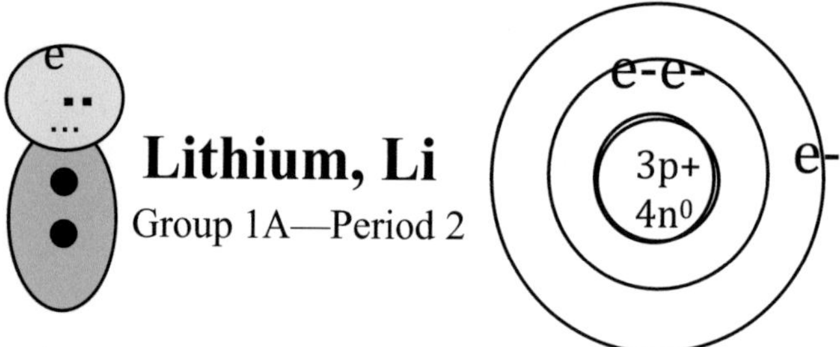

BERYLLIUM

Leaving Lithium, they headed back to mulberry bushes where they left their magic flying bubble. Before taking off. Professor Terry said, "It's time to look at our map. Let's see which element we'll visit next. We just saw element Atomic Number 3. Next will be element Atomic Number 4, Beryllium. Let's color in Lithium now that we have visited him. He's green because that's the color for elements in Group 1A."

The trip to 2nd Street went fast. It was a short distance, and they were soon there.

PERIODS ↓ GROUPS →

MAP of PERIODIC TABLE LAND

	1A	2A		3A	4A	5A	6A	7A	8A
1.	1 **H** 1.00								2 **He** 4.00
2	3 Li 6.94	4 **Be** 4.01	B GROUPS REMOVED	5 **B** 10.8	6 **C** 12.01	7 **N** 14.01	8 **O** 16.99	9 **F** 18.99	10 **Ne** 20.18
3	11 **Na** 22.9	12 **Mg** 24.3		13 Al 26.96	14 Si 28.09	15 **P** 30.97	16 **S** **32.07**	17 **Cl** 35.45	18 Ar 39.95
4.	19 **K** 39.1	20 Ca 40.08							

4 **Be** 9.01

e-e-

Beryllium, Be, Atomic #4.
Group 2A—Period 2

They landed the balloon in the cleared area behind the trees on 2nd Street and made their way down to #4, 2nd Street the home of Beryllium. There was Beryllium standing by the gate looking happy to see them. Professor Terry announced, "Guy is going to draw your Bohr model. When he finishes your Bohr model, you'll soon see that your atom will confirm the meaning of your body markings as well as all that information in your Element Box on the Periodic Table."

Beryllium said enthusiastically, "I can't wait to learn how to draw my Bohr model." Guy showed him the Bohr model that he had drawn for Hydrogen, and explained that his would look different since every element's atom is different. Guy had brought along the chart that he had made to show the steps he would follow in drawing the Bohr model. Guy said, "If you follow this chart as you watch what I'm doing, you will be able to draw your own Bohr model some day."

After touring the house, the room that looked out over the garden was the place Beryllium had set up to draw his Bohr model. On the table by the window there was a large pad ready for Guy to use. Next to that table he found a perfect place to hang the Steps Chart, and Guy began. He knew that Beryllium would be happy to see his atom.

Guy Gets Busy Drawing Beryllium's Bohr Model

1. Label the columns:

ELEMENT BOX CALCULATIONS BOHR MODEL

2. Copy and Label the Element Box from Periodic Table.

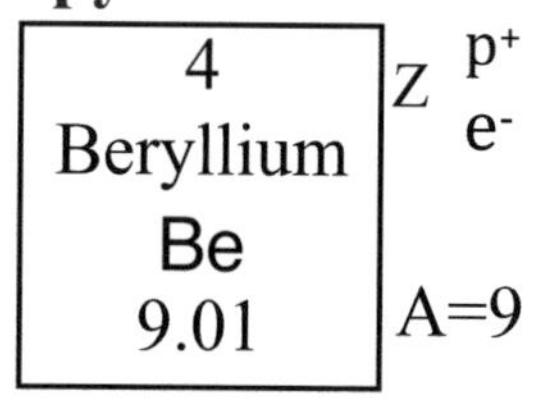

3. Do the Calculations

$n^0 = A - Z$

$n^0 = 9 - 4$

$n^0 = 5$

4. Draw Beryllium's Atom

1st **Draw** a thick Circle for the nucleus.

Beryllium's Atom

2nd **Draw Energy Levels**. Beryllium's in Period 2 . So, draw 2 energy levels.

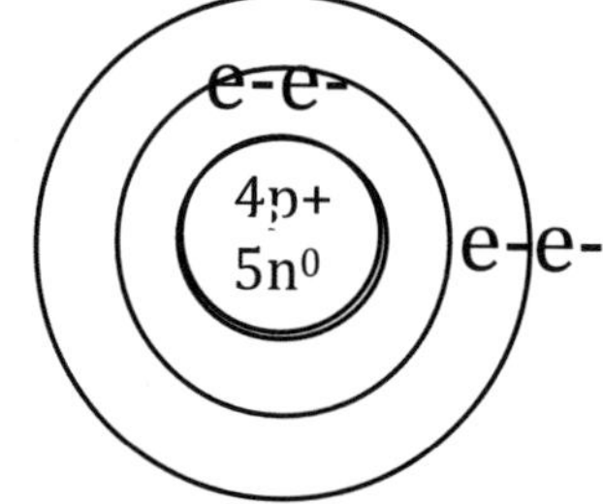

3rd Put the Protons and Neutrons in the nucleus.

p+= Atomic # Beryllium's is 4

n^0 neutrons (See Calculations).

4p+
$5n^0$

4th Put electrons in Energy Levels. How many?
It equals the Atomic Number.
Beryllium's Atomic Number is 4.
So put 4 electrons in the energy levels.

Guy said, "I've finished drawing your Bohr model. Come look at it."

...

BERYLLIUM'S BOHR MODEL

ELEMENT BOX | **CALCULATIONS** | **BOHR MODEL**

4
Beryllium
Be
9.01
Z p^+ e^-
A=9

$n^0 = A - Z$

$n^0 = 9 - 4$

$n^0 = 5$

e-e-

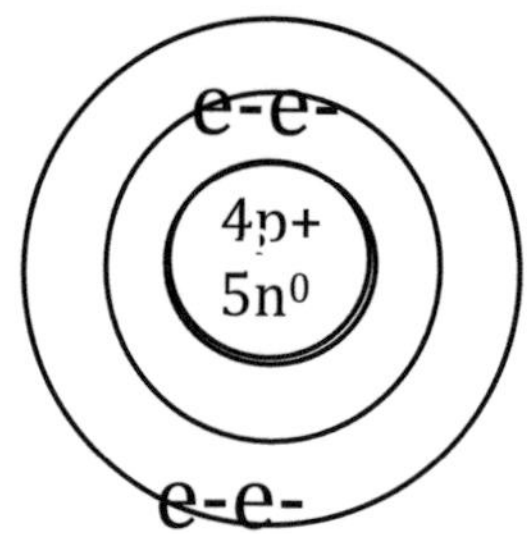

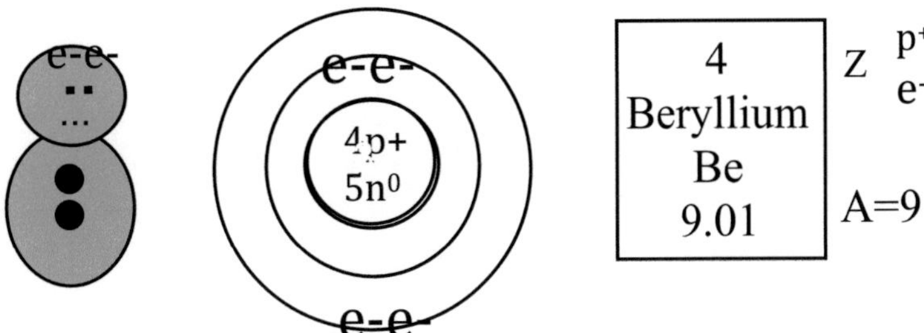

Beryllium looked at his Bohr model and said, "Now my element box means so much more to me. I can actually see that my **Atomic Number** 4 equals the 4 protons I have in my nucleus and the 4 electrons I have running around my nucleus in my energy levels. The Bohr model also shows me which energy levels these electrons are in. This also helps me understand better the meaning of being in **Period 2** on the Periodic Table. I always knew it meant I had 2 energy levels, but now I can see I really do have 2 energy levels in my atom. I can also see why I need 2 energy levels. I have `4 electrons, and the first energy level next to the nucleus can only hold 2 electrons. So I need an energy level for my other 2 electrons. It all makes sense to me now. My body markings make sense too. My 2e hair are those 2 electrons in my Outside Energy Level which also means I'm in Group 2A. The 2 dots on my body represent my 2 energy levels. Wow! Now I'd like to tell you how people have learned to use my element. Professor Terry and Guy were happy to learn something about Beryllium. Beryllium explained, "When I talk about my uses, it sometimes means I'm talking about one of my compounds." So they went into a room with comfortable chairs and sat down and Beryllium began.

BERYLLIUM

Beryllium said, "You'll soon know more about me. I'm one of the lightest metals in the universe, so they use me in high performance aircraft and in space vehicles. I fly high and see the Earth like few people have ever seen it. The Earth is called the Blue Planet, and now I know why. From space it looks like a blue globe. It's also true that I'm a hard metal, but I'm brittle at room temperature. **Beryllium** salts have a sweet taste like sugar, glucose. Because of this, Beryllium was once called glucine." Beryllium continued, "I have a long history. I was even known to the ancient Egyptians who valued me in the gem stone beryl. Today I'm valued in a very precious green gem the emerald."

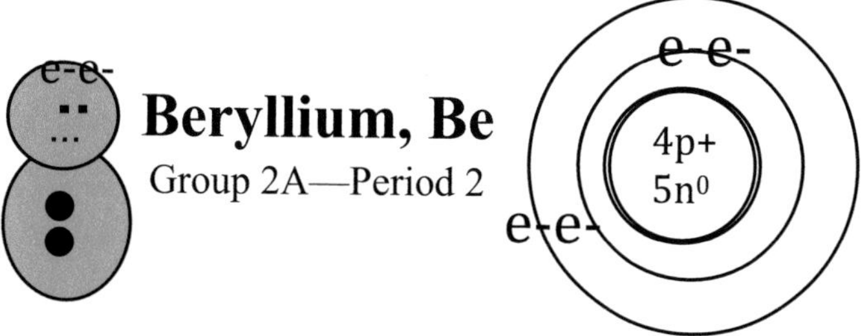

BORON

The trip from 2nd Street to 3rd Street took a long time as they had to fly again over the wide B Avenue covering the whole central portion of Periodic Table Land. Guy loved the different houses there. Guy loved Silver's home. It was a castle with seven spires each one glistening in the sun and sending sparkling reflected light off in every different direction. Gold had a smooth golden dome. Iron's home was in the shape of a black frying pan with windows all around.

Professor Terry said, "Let's take a look at our map. We completed our visit to Beryllium. Let's color in Beryllium's element box brown. That's the color of all the elements in Group 2A. Later you'll learn that this is their family color

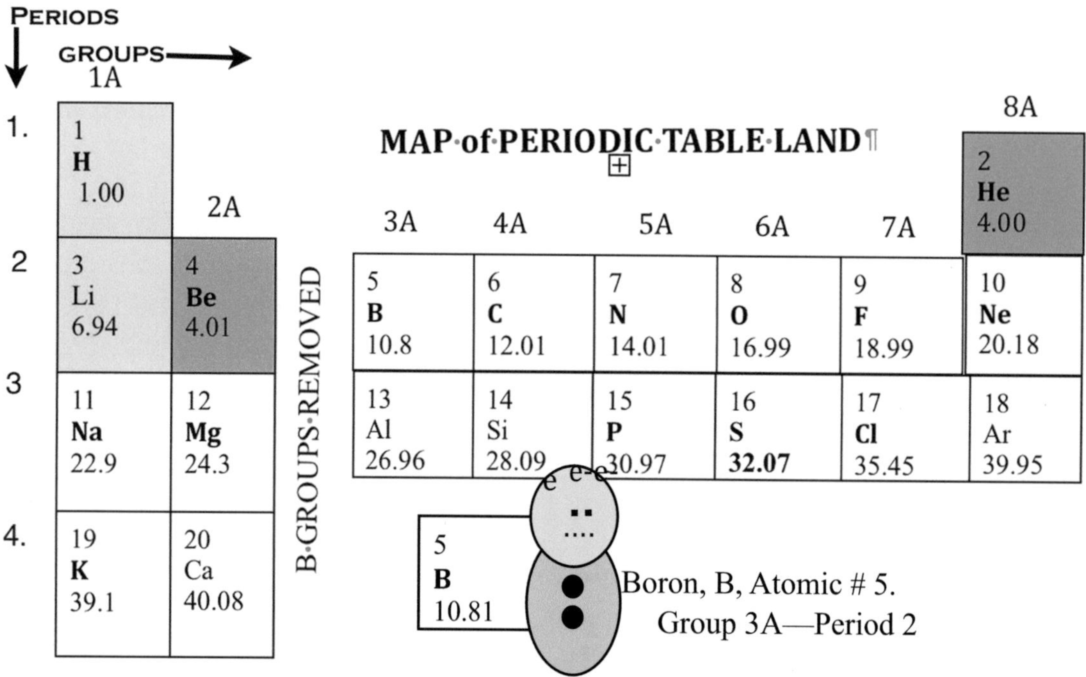

Soon they landed the magic bubble on the tarmac behind 3rd Street. They folded away the balloons in a safe place behind some flowering lilac bushes before they made their way down to Boron's home at # 5, 3rd Street. Guy loved the sweet smelling purple flowers

When they got there they knocked and soon the door opened. **Boron** smiled and welcomed them. Guy explained that he had come to draw his Bohr model. Boron had been interested in seeing what his atom looked like for a long time.

Guy said, "The Bohr model will confirm the meaning of the markings on your body as well as explaining your Element Box and the meaning of your Group and Period on the Periodic Table. You'll see." Guy showed Boron Hydrogen's Bohr model saying, "Your atom will be similar to Hydrogen's but uniquely your own. Each element's atom is totally different from the atoms of every other element in the universe. He also showed Boron the Steps Chart, so he could follow along and learn how it's done. Guy began

Guy Gets Busy Drawing Boron's Bohr Model

1. Label the columns:

ELEMENT BOX CALCULATIONS BOHR MODEL

2. Copy and Label the Element Box from the Periodic Table.

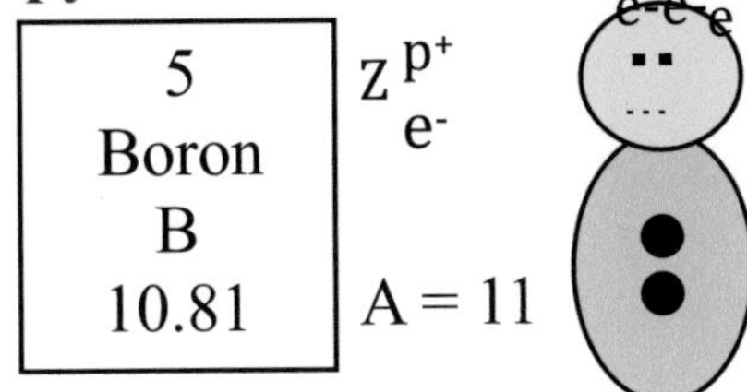

3. Do the Calculations

$n^0 = A - Z$

$n^0 = 11 - 5$

$n^0 = 6$

4. Draw Boron's Atom

1st Draw a thick circle for the nucleus

2nd Draw Energy Levels. Boron is in Period 2 . So, draw 2 energy levels.

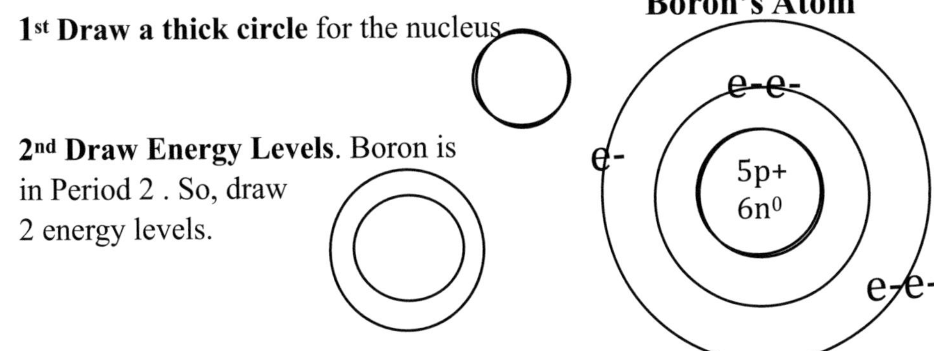

3rd Put the Protons and Neutrons in the nucleus.

p+= Atomic # Boron's is 5 5p+

n^0 neutrons (See Calculations). $6n^0$

4th Put electrons in Energy Levels. How many?
It equals the Atomic Number.
Boron's Atomic Number is 5.
So put 5 electrons in the energy levels.

Guy said, "I've finished drawing your Bohr model.
Come look at it."

BORON'S BOHR MODEL

ELEMENT BOX CALCULATIONS BOHR MODEL

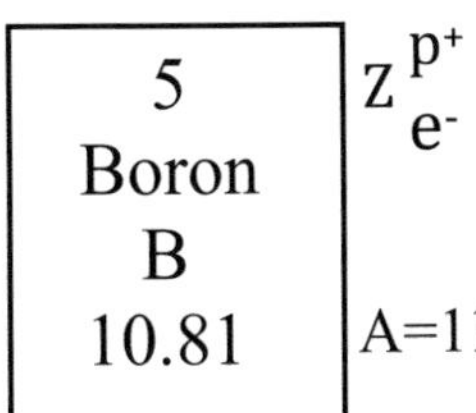

$n^0 = A - Z$

$n^0 = 11 - 5$

$n^0 = 6$

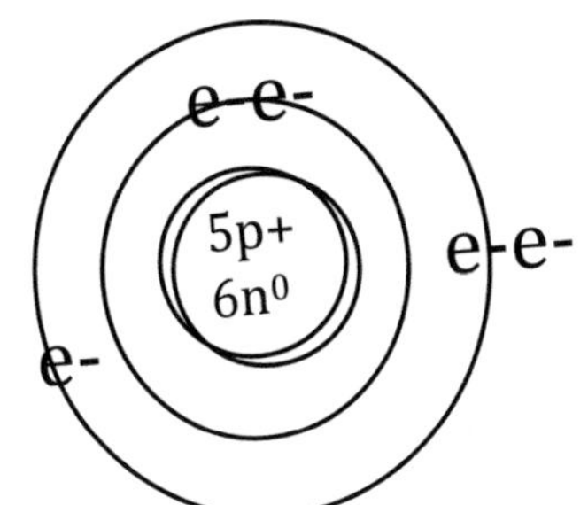

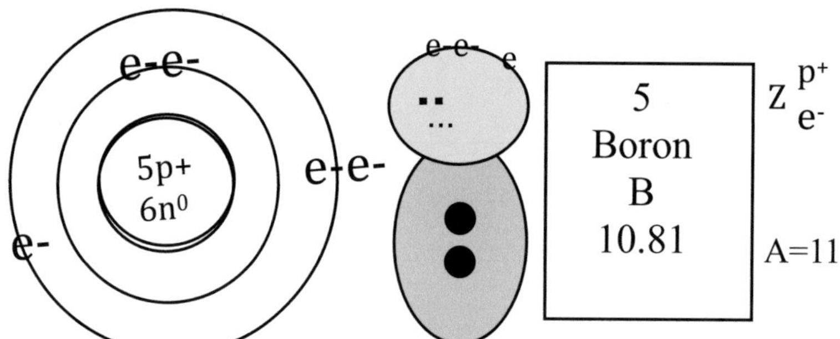

Boron totally amazed said, "Wow! My element box really does represent my atom. My Atomic Number is 5, and there are 5 protons in my nucleus and the 5 electrons in my energy levels. This model of my atom also makes me understand the meaning of being in Period 2 on the Periodic Table. I always knew it meant I had 2 energy levels but now I can see I really do have 2 energy levels. Count them. The model of my atom also explains my body markings. The 3 electrons in the Outside Energy Level tell that I'm in Group 3A and why I have 3e hair. The 2 dots on my body mean I have 2 energy levels. You can see them in my Bohr model This is great!. It all makes sense to me now."

"I'd like to tell you a little about me." Professor Terry and Guy went into Boron's house and sat down eager to learn about Boron. Guy was so happy to learn more about another element that's part of his magnificent world.

BORON

"I've been used for years in the product, borax. It is known for making your wash come out whiter." Boron continued, "My uses are mainly made with compounds of Boron. I'm used in fiberglass to make it insulate better, and in structural materials to make them stronger. I have a wide variety of uses. I give flares a green color. I'm a rocket fuel igniter. I'm found in eye drops, mild antiseptics and tile glazes. I'm also used in making beautiful, decorative ceramics that people love. Boron compounds play a strengthening role in plant walls. borate compounds and boric acid can also be used as a pesticide."

Professor Terry and Guy learned a great deal about Boron as they sat listening intently. When Boron finished it was time to move on.

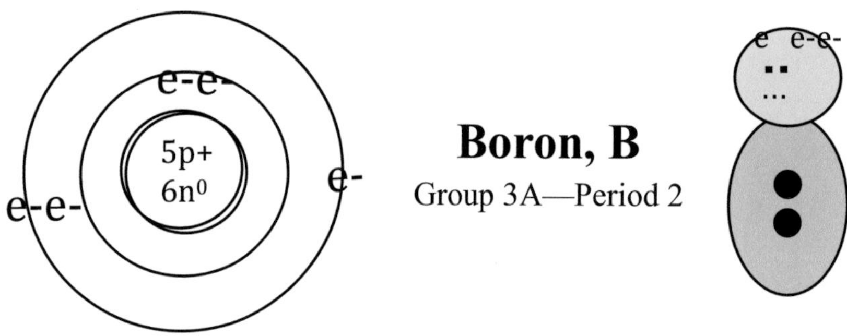

CARBON

Before they climbed into the magic bubble Professor Terry reminded Guy that they needed to color in the Element Box for Boron. Boron was right there on the map in Group 3A, Period 2. Guy colored the box light grey, Boron's family color.

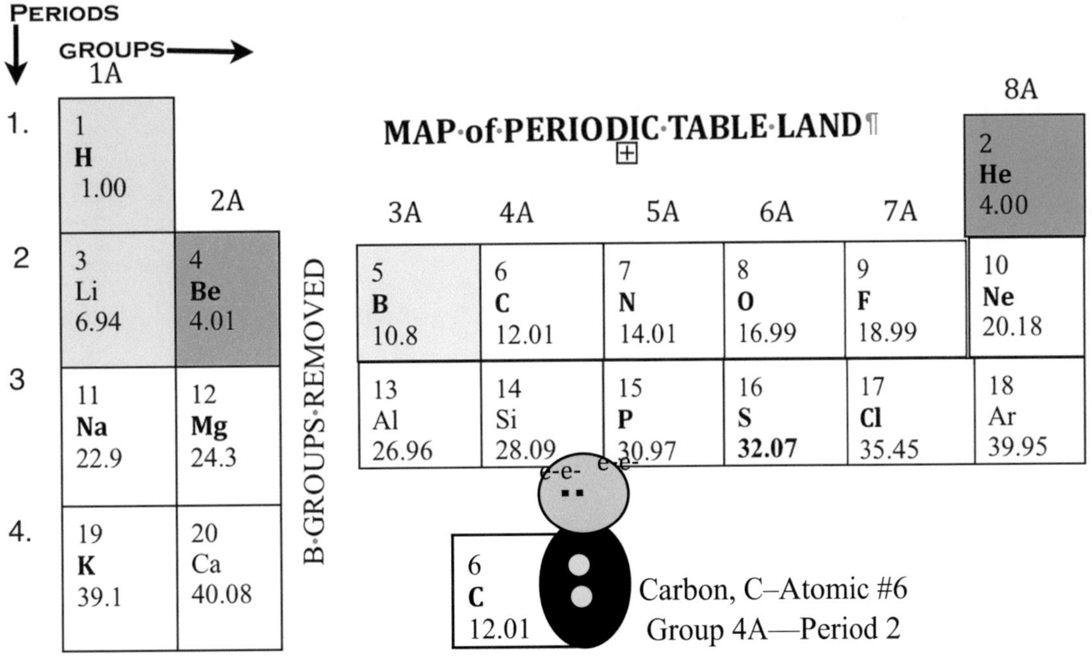

The magic bubble took off, following the direction Guy pointed with the magic wand. Up, up and away it rose giving them wonderful views of the adorable houses below. Carbon's log cabin soon came into view. Carbon is the main element found in wood which would explain Carbon's decision to have a log cabin for his home. Passing over Carbon's home they landed in that smooth grassy field at the top of 4th Street. Guy was hoping to be invited to see inside Carbon's log cabin.

After knocking, Carbon quickly opened the door and said, "I heard you're going around teaching all the elements how to draw their Bohr models. News travels fast around here. Come on in." Observing Guy's interest in the house Carbon gave him a tour. It was fascinating. All the furniture was made of hand carved wood. To make the seats comfortable there were soft and colorful cushions and fluffy pillows on all the chairs. It was a most interesting house.

Then, they were back discussing the Bohr model that Guy had come to draw. "Look at this model we drew for Hydrogen. Your atom will be very different because each element's atom is distinctly different from all others. That's what makes elements special. Most important, when you see your atom you will understand better the meaning of your Element Box on the Periodic Table and your body markings."

Guy showed Carbon the Steps Chart and suggested reading it while Guy drew his atom. "It will help you draw it yourself some day." Then Carbon picked a spot to hang the Steps Chart. Guy picked up Carbon's tablet and began to draw Carbon's Bohr model.

Guy Gets Busy Drawing Carbon's Bohr Model

1. Label the columns:

ELEMENT BOX CALCULATIONS BOHR MODEL

2. Copy and Label the Element Box from the Periodic Table.

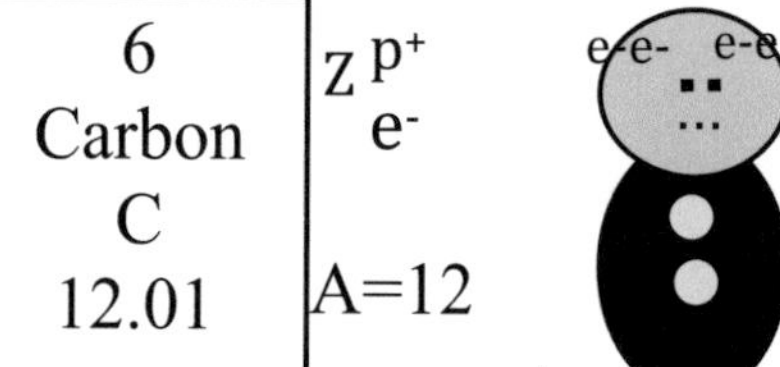

3, Do the Calculations

$n^0 = A - Z$

$n^0 = 12 - 6$

$n^0 = 6$

4. Draw Carbon's Atom

1st draw a thick circle for the nucleus.

2nd Carbon is in **Period 2.**
So, Draw 2 Energy Levels around the nucleus.

Carbon's Atom

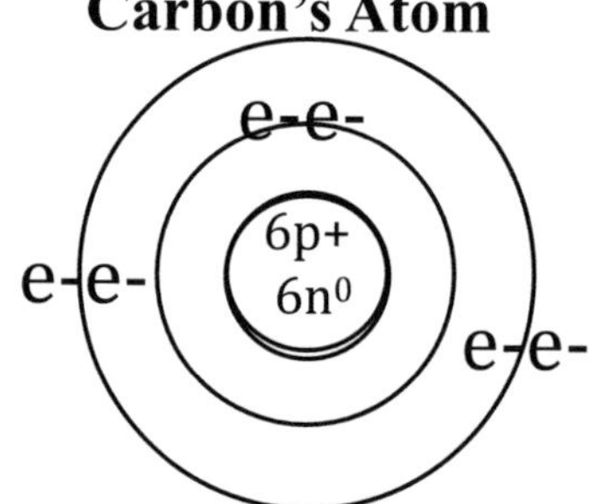

3rd Put the protons and neutrons in the nucleus.

p+ = Atomic # Carbon's is 6 6p+

n^0 neutrons (See Calculations). $6n^0$

4th Put electrons in Energy Levels.

# of electrons equals the Atomic#.

Carbon's Atomic Number is 6.

Put 6 electrons in the energy levels.

Guy said, "I've finished drawing your Bohr model. Come look at it."

CARBON'S BORH MODEL

ELEMENT BOX		CALCULATIONS	BOHR MODEL
6 Carbon C 12.01	Z p+ e- A=12	$n^0 = A - Z$ $n^0 = 12 - 6$ $n^0 = 6$	

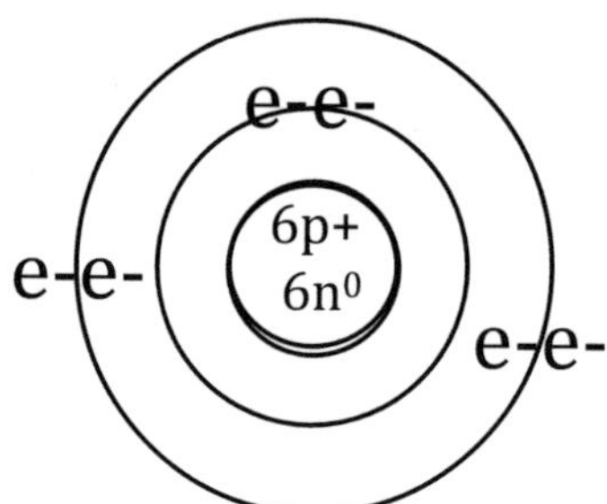

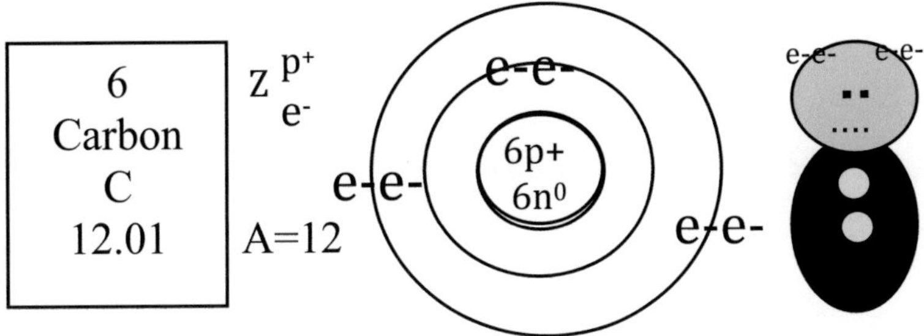

Carbon looked at his Bohr model and said, "Now I look at my Element Box and the numbers take on meaning. I see that my Atomic Number represents the 6 protons I have in my nucleus and the 6 electrons I have running around my nucleus in their energy levels. The Bohr model also shows me which energy levels these electrons are in. This also makes me understand so much better the meaning of being in Period 2 on the Periodic Table. I always knew it meant I had 2 energy levels, but now my Bohr model shows me these 2 energy levels. The 4 electrons I have in my Outside Energy Level tells me I'm in Group 4A, and explains my 4e hair. My Bohr model tells me so much!"

Carbon was delighted to see his Bohr model and said, "Now that you've finished drawing and explaining my Bohr model, I would like to have you sit down and visit with me. I'd like to tell you about myself." That meant Guy would gather more information about the world he loved so much.

CARBON

Carbon said, "I'm a very different kind of element. I'll just tell you some of my more familiar compounds: sugar, butter, gasoline, propane, octane, ether, alcohol. We breathe Carbon Dioxide out of our lungs. The leaves of plants take in carbon dioxide and use it in photosynthesis which puts out Oxygen into our air. Carbon is found in all living things. Coal and diamonds are both pure carbon. Yet they look nothing alike. Coal is used to heat houses. Pure crystal diamonds are used in rings and other kinds of jewelry." Carbon continued, "My element's most important feature is that a bunch of us Carbon atoms can form long chains by joining our atoms together. That's the reason we can make so many different kinds of compounds."

Guy was absolutely fascinated. He thought, "That's another explanation for how there can be so many different things in the world when there are only 92 elements in nature contributing to this diversity. Wish Star was explaining this to me."

Professor Terry thanked Carbon for sharing all that information with them. Then off they went knowing so much about Carbon and learning what his atom looked like.

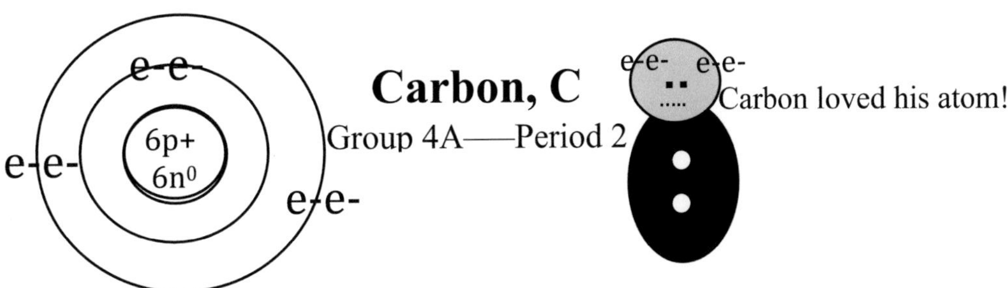

NITROGEN

Guy and Professor Terry returned to the grassy field and found the bubble with its lovely green balloons where they left them. Professor Terry said, “Let’s mark the map.”

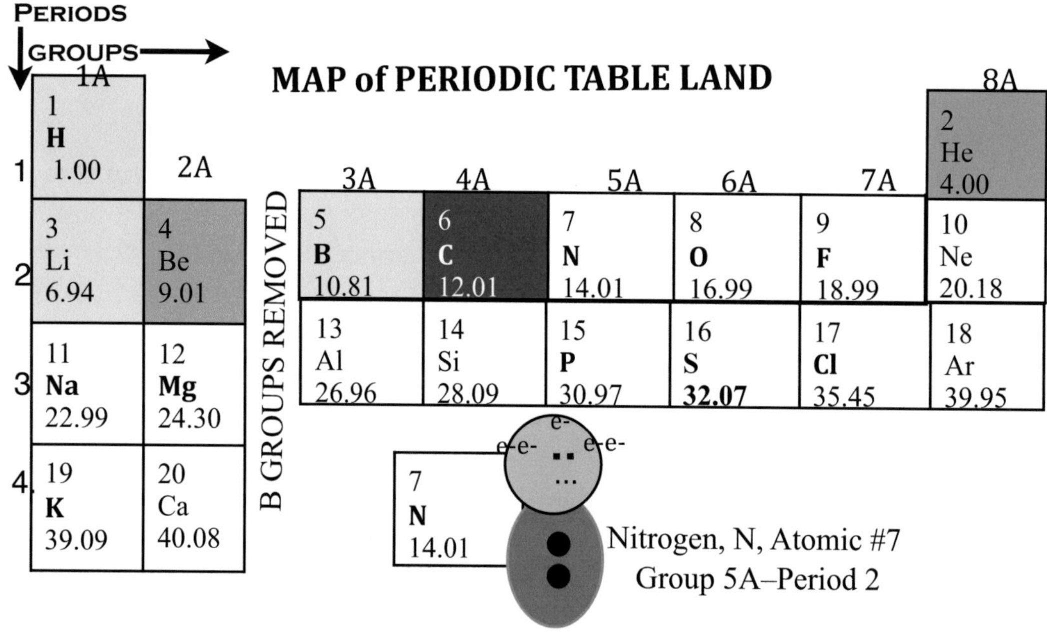

Guy looked at the map and found Carbon, the element that they had just visited. He colored in Carbon’s element box black, Carbon’s family color. “I see on the map that we’re going to visit Nitrogen next.” They jumped into the bubble and off they flew.

In no time they had landed in the area behind 5th Street. Finding Nitrogen was next on the list. He was located at #7, 5th Street. The closer they got to Nitrogen’s home the more beautiful the flowers appeared alongside the road. When they peeked over the gate into Nitrogen’s garden, they saw the most magnificent display of flowers that they had ever seen. They spied Nitrogen in the side garden pulling weeds. “Hi Nitrogen,” called Guy, “I’ve come to draw your Bohr model.” Nitrogen came over to the gate, lifted the latch and invited them in. “I was told to expect you. I really would like to see what my atom actually looks like.”

Guy responded, “It’s more than just seeing what your atom looks like. When you see your Bohr model, you will better understand your Element Box on the Periodic Table. You will also appreciate more the meaning of your 5 e hair and the 2 dots on your chest.”

“I created a chart that I call a Steps Chart. It shows step by step how to draw your Bohr model. I made it to help you follow along as I draw your Bohr model. I’d like to hang the Step Chart where you’ll have a good view of it as well as a view of the diagram of your atom that I’m drawing. If you follow along with me, I feel that later you will be able to draw your own Bohr model without any help.”

Nitrogen said, “Let’s go out in the rose garden. That way we can all enjoy the beautiful flowers while working on my Bohr model.” Guy was happy to be out in nature and he picked up the marker and began drawing.

Guy Gets Busy Drawing Nitrogen's Bohr Model

1. Label the columns:

ELEMENT BOX CALCULATIONS BOHR MODEL

2. Copy and Label the Element Box from the Periodic Table.

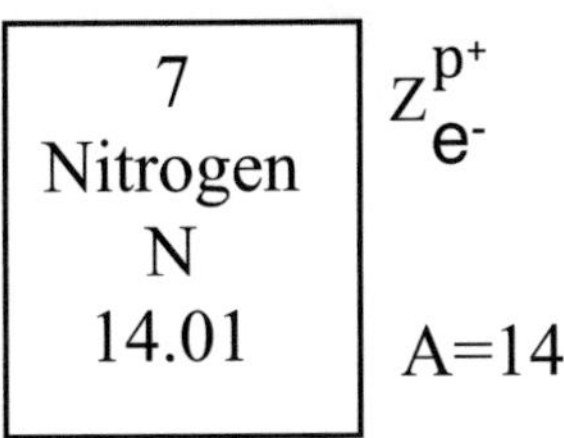

3. Do the Calculations

$n^0 = A - Z$

$n^0 = 14 - 7$

$n^0 = 7$

4. Draw Nitrogen's Atom

1st draw a thick circle for the nucleus.

Nitrogen's Atom

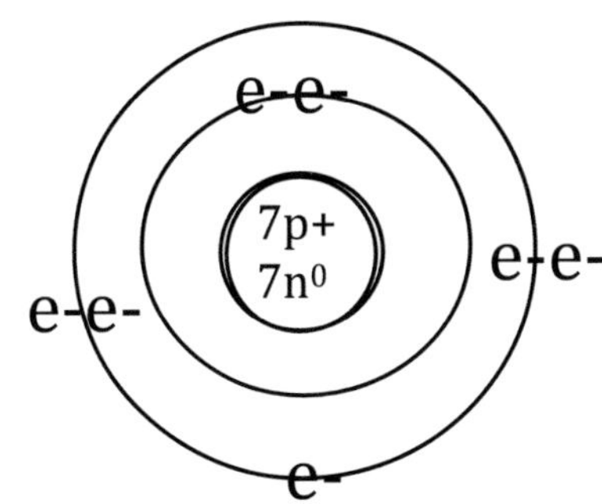

2nd Nitrogen is in **Period 2.**
So, Draw 2 Energy Levels around the nucleus.

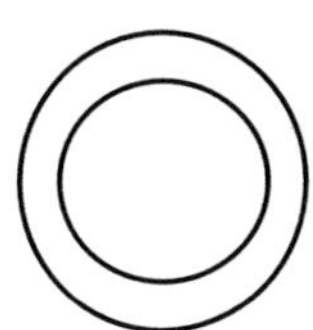

3rd Put the protons and neutrons in the nucleus.

p+ = Atomic # Nitrogen's is 7 7p+

n^0 neutrons (See Calculations). $7n^0$

4th Put electrons in Energy Levels.

# of Electrons equals the Atomic Number.
Nitrogen's Atomic Number is 7
Put 7 electrons in the energy levels.

Guy said, "I've finished drawing your Bohr model. Come look at it."

NITROGEN'S BOHR MODEL

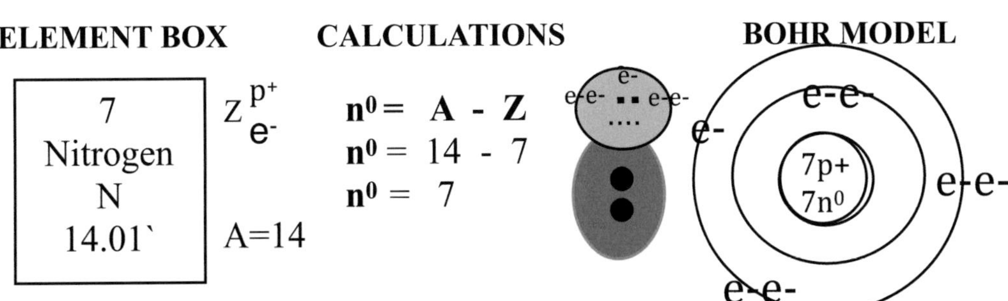

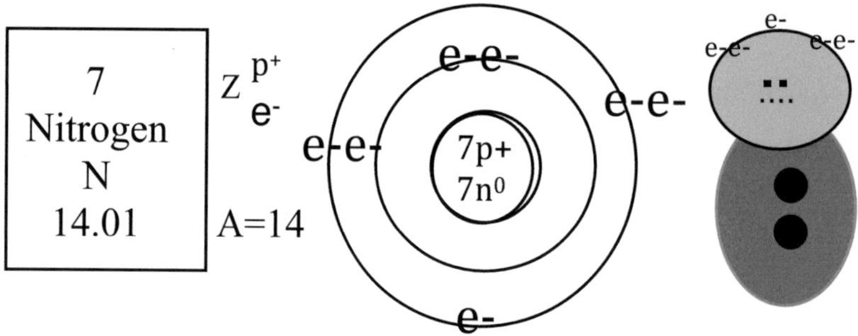

Nitrogen said, "My Bohr model makes my Element Box come alive. Now when I look at the Atomic Number 7 in my Element Box, I see the 7 electrons moving in the energy levels, and the 7 protons inside my nucleus. When I see Period 2 on the Periodic Table and the 2 dots on my body I see the 2 energy levels in my atom. I know I'm in Group 5A on the Periodic Table. Now I actually see 5 electrons in my atom's Outside Energy Level, and it explains my 5e hair. Amazing!" Nitrogen was delighted to see his Bohr model and said, "Now that you've finished drawing and explaining my Bohr model, I would like to have you sit down and visit with me."

NITROGEN

The colorful and sweet smelling garden was a nice setting to hear all about Nitrogen. Nitrogen began, "My nitrate compounds are made into fertilizers. My garden grows so well because I use these fertilizers. Farmers need to fertilize the soil where the crops grow to keep the crops healthy. They replace the nitrogen in the soil using a process called *rotation of crops.* Each year they plant clover on one quarter of the land. Clover puts nitrogen back into the soil. Now let's think about the air we breathe. Everyone thinks of air as oxygen because that's the part of air people need to breathe. Nitrogen makes up 78% of the air you breathe. That's a large part of the air." Nitrogen continued, "When you see protein on a food label, know that this food contains Nitrogen. I'm there to make you strong. I'm even in the DNA that is part of what makes you, you. I'm an important part of life." With that Nitrogen was finished. Professor Terry and Guy thanked Nitrogen and off they went to see Oxygen. Then Nitrogen had time to enjoy looking at his Bohr model.

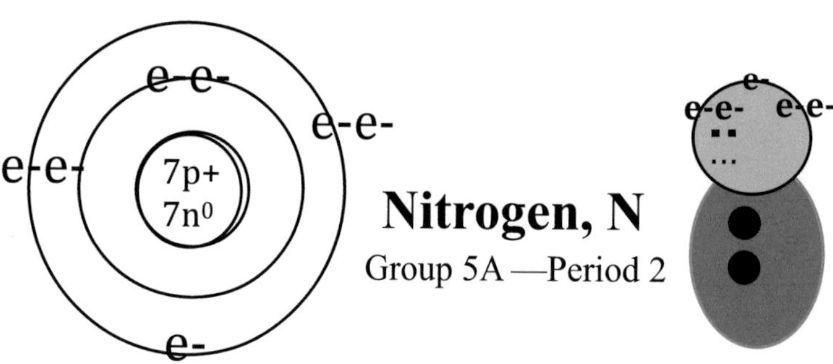

OXYGEN

Once again, the balloons lifted the magic bubble and off they flew high up above the element's adorable houses. The sky was bluer today. There were just a few fluffy clouds in the distance, far from 6th Street where Oxygen lived. As the balloon descended Professor Terry noticed there was a small airport at the head of 6th Street. "Guy, check out the hangar at end of the airport. It looks like a safe place to keep our bubble and green balloons." After landing Professor Terry said, "Wait a minute Guy we forgot to do something. We have to look at the map and color in the box for the element we just visited."

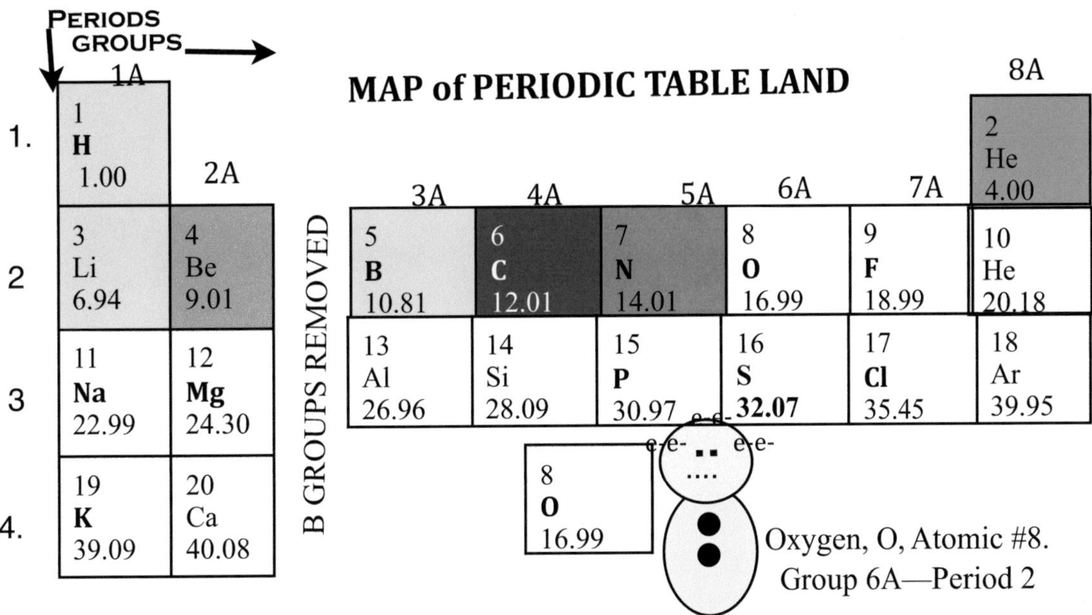

Guy remembered. "That was Nitrogen in Group 5A Period 2." Guy colored the box red because that is Nitrogen's family color. This finished, they quickly made their way down to #8, 6th Street, Oxygen's home. When they got there they knocked at the door. Soon the door opened, and Oxygen stood smiling at them.

Guy explained, "I'm going around Periodic Table Land drawing Bohr models for all the elements. You are next on the list of elements to visit."

Oxygen said, "I heard what you have been doing, and I've always wanted to see what my atom looks like. So come on in. They entered and found the air so fresh inside Oxygen's house. Oxygen is the part of the air our lungs need to keep us alive."

They sat down at the kitchen table and had a refreshing glass of cold water. Professor Terry said, "When you see the model of your atom, you will be surprised at all the information it will explain. You will learn so much about your body markings and your Element Box."

Guy said, "I made a chart for you that tells step by step how to draw a Bohr model. I'd like you to be able to follow along on the chart what I'm doing as I draw your Bohr model." They found that perfect spot to hang the Steps Chart. It was right next to the table that Oxygen had set up. On the table was a large pad and a marker. Guy began.

Guy Gets Busy Drawing Oxygen's Bohr Model

1. Label the Columns:

ELEMENT BOX **CALCULATIONS** **BOHR MODEL**

2. Copy and Label the Element Box from the Periodic Table.

8 Oxygen O 16.99

Z $^{p^+}_{e^-}$

A =17

e-e-
e-e- ▪▪ e-e-
...

3. Do the Calculations

$n^0 = A - Z$
$n^0 = 17 - 8$
$n^0 = 9$

4. Draw Oxygen's Atom

1st draw a thick circle for the nucleus.

2nd Oxygen is in **Period 2.** So, **Draw** 2 Energy Levels around the nucleus

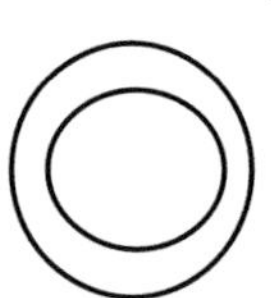

Oxygen's Atom

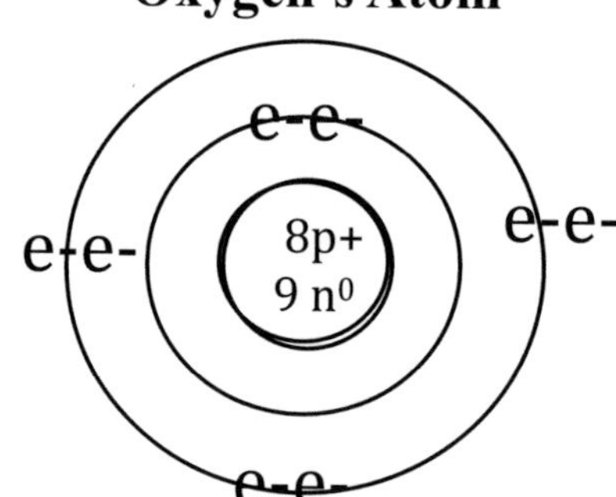

3rd Put the protons and neutrons in the nucleus.
p+ = Atomic # –Oxygen's is 8. 8p+
n^0neutrons (See Calculations). 9 n^0

4th Put electrons in Energy Levels.
# of Electrons equals the Atomic Number.
Oxygen's Atomic Number is 8
Put 8 electrons in the energy levels.

Guy said, "I've finished drawing your Bohr model. Come look at it."

OXYGEN'S BOHR MODEL

ELEMENT BOX **CALCULATIONS** **BOHR MODEL**

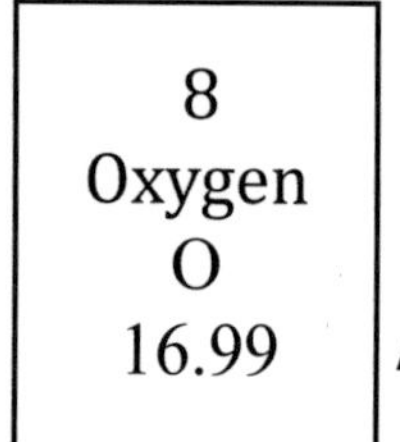

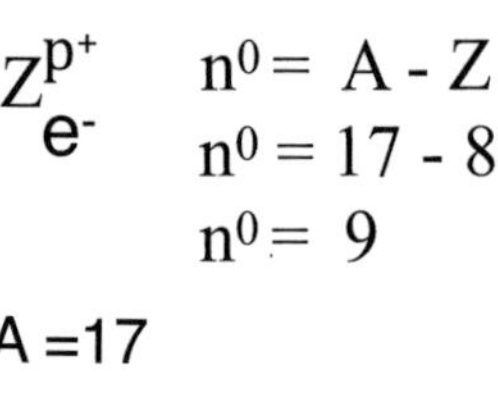

e-e-
e-e- ▪▪ e-e-
...

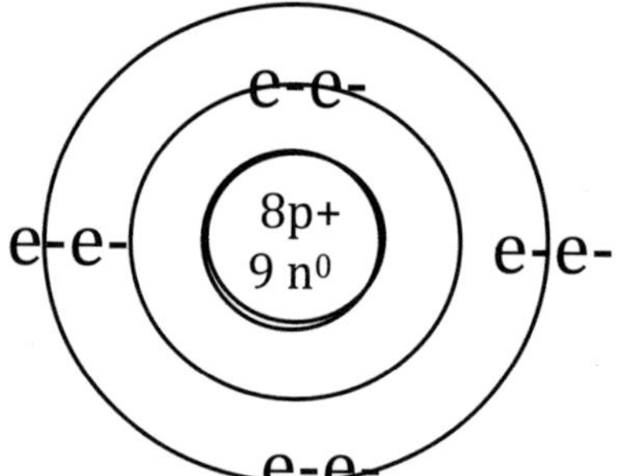

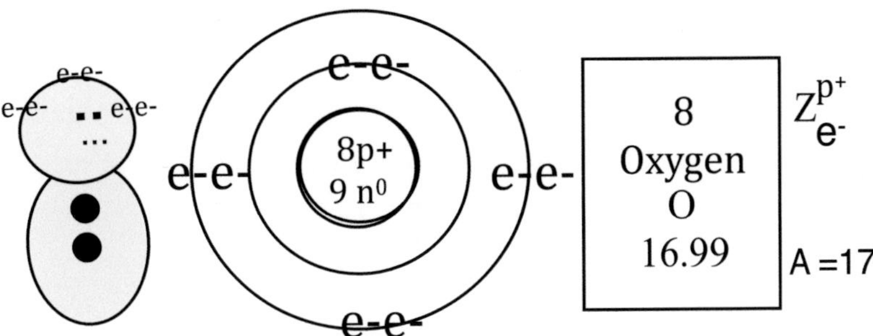

Oxygen was delighted to see his Bohr model. Next he decided to check it out. "My Atomic Number in my Element Box is 8. Checking, my atom has 8 protons in the nucleus and 8 electrons in my energy levels. I'm in Period 2 on the Periodic Table. Checking. My atom shows I have 2 energy levels which is also what the 2 dots on my body have always told people. It's exactly as it should be. In my atom there are 6 electrons in my atom's Outside Energy Level meaning I'm in Group 6A. That's what my 6e hair means . Wow! My Bohr model is great."

Oxygen invited Guy and Professor Terry to stay and visit.

OXYGEN

Oxygen said, "I have a lot to tell you about my element and compounds that I am part of. I'm a colorless, odorless gas, and I make up 20% of our air. Leaves on trees create Oxygen and send it out in the air for people to breathe." Oxygen continued, "When I form compounds with metals, my compounds are called Oxides. When I combine with Iron, I form a compound called Iron oxide which is rust. That makes Iron very weak. I'm not too popular when I do this. I am found combined with Hydrogen as the compound water, H_2O. Water is absolutely necessary for life. Living things can not live without having me around to breathe. When you study polyatomic ions, you will see how I'm part of Nitrates, Phosphates, Carbohydrates, Carbonates and Hydroxides. Oxygen is found in many things".

Professor Terry and Guy thanked Oxygen for sharing so much about his element and the compounds he forms.

Oxygen looked, and looked, and just loved his Bohr model.

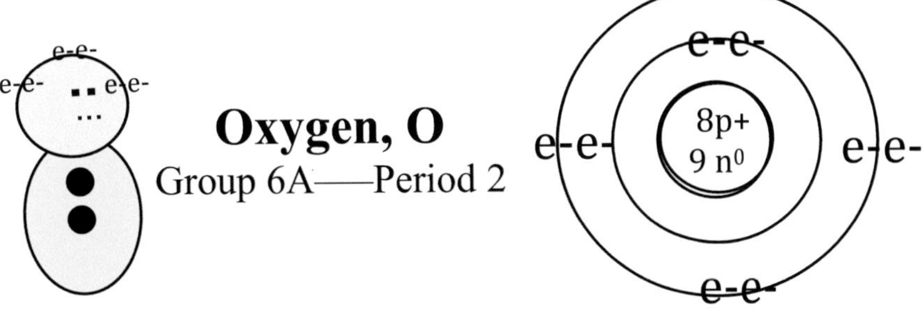

FLUORINE

Since they were presently on 6th Street visiting Oxygen, it wouldn't take too long to get to Fluorine on 7th Street. It seemed they got there in no time at all. As soon as they landed Guy remembered to take out the map and color in Oxygen's box. Professor Terry said to make it yellow because that is the Oxygen family's color.

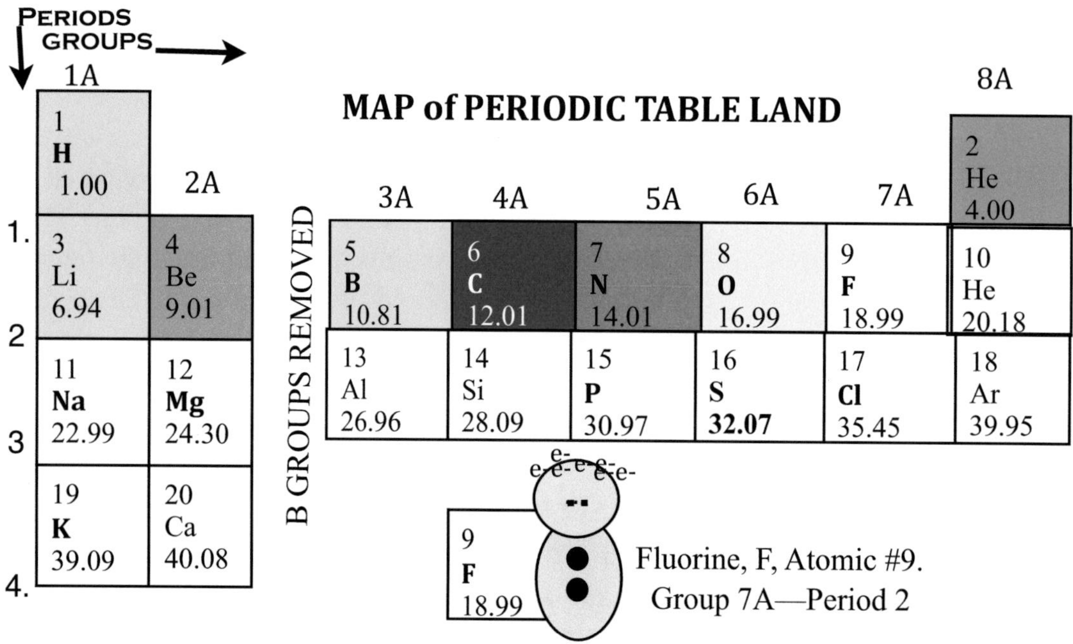

As they emerged from the bubble, the aroma of pine that Guy loved so much met them. Behind 7th Street was a thick Pine forest. Professor Terry said, "The first thing to do is to stash the balloons and bubble in a safe place. There's a good spot over here behind these bushes underneath this really old pine tree."

Guy remarked, "Look how thick the tree's trunk is. It must be very old." They saw so many pine trees that it was obvious why this was called the Pine Forest. When they emerged from the forest, they saw Fluorine in a very strange place.

He was standing on the roof of his house. Actually the house was a strange shape too. It was in the shape of a tooth. Professor Terry told Guy that some compounds of Fluorine were used to prevent tooth decay. That explains why he chose that shape for his house. Fluorine invited them to climb up the ladder to join him on the roof. They did.

Fluorine said, "I'm ready for you. I heard all about your adventures drawing Bohr models. Here's a stand that you can hook your Steps Chart on. I'm prepared for you to draw my Bohr model. I can't wait to see what it looks like."

Guy said, "So you even heard about the Steps Chart. That is a good place to hang it. Did they tell you how exciting it is to see that the Bohr model really displays the information from the Periodic Table and your body markings? You'll see. The Steps Chart tells step by step how to draw a Bohr model. I'd like you to look at it as I draw your atom. That's a good size pad you have for me to draw on. Let's get started."

Guy Gets Busy Drawing Fluorine's Bohr Model

1. Label the Columns:

ELEMENT BOX CALCULATIONS BOHR MODEL

2. Copy and Label the Element Box from the Periodic Table.

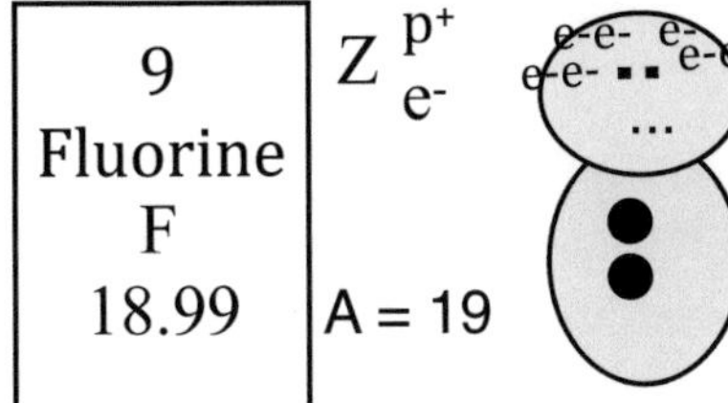

3. Do the Calculations

$n^0 = A - Z$

$n^0 = 19 - 9$

$n^0 = 10$

4. Draw Fluorine's Atom

1st draw a thick circle for the nucleus

2nd Fluorine is in **Period 2.**
So, **Draw** 2 Energy Levels around the nucleus.

3rd Put the protons and neutrons in the nucleus.
p+ = Atomic # – Fluorine's is 9
n^0neutrons (See Calculations).

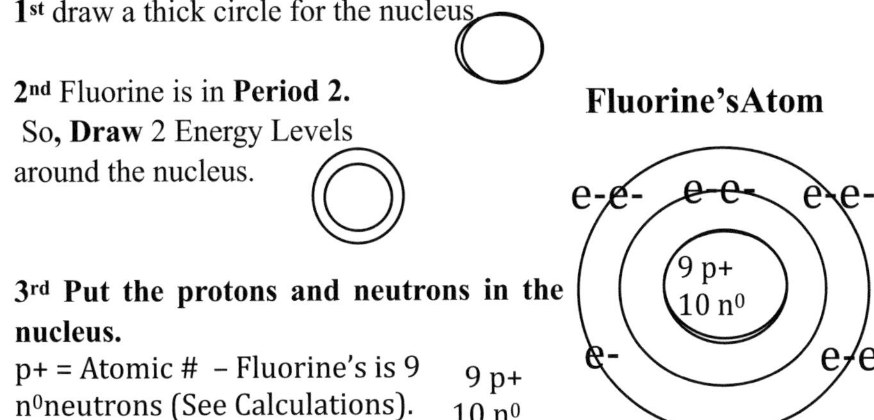

4th Put electrons in Energy Levels.
# of Electrons equals the Atomic Number.
Fluorine's Atomic Number is 9
Put 9 electrons in the energy levels.

Guy said, "I've finished drawing your Bohr model. Come look at it."

FLUORINE'S BOHR MODEL

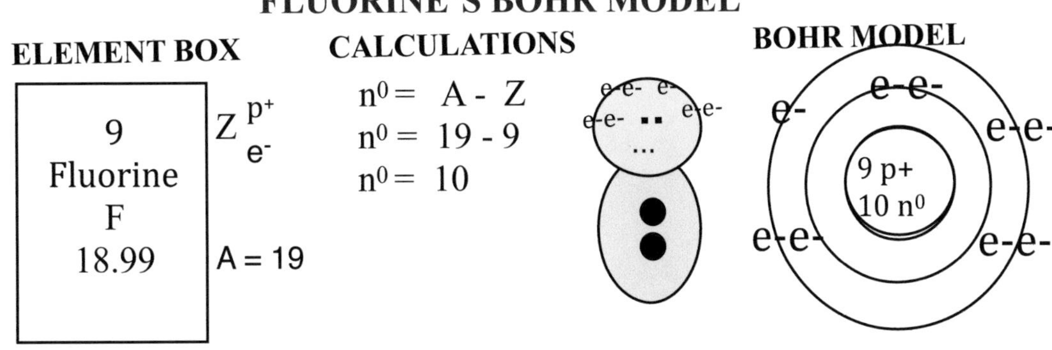

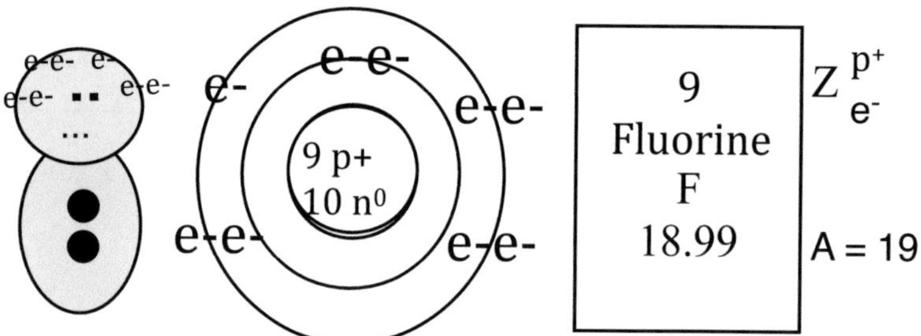

Fluorine said, “My Bohr model tells me so much about myself. The 7 electrons in my atom’s Outside Energy Level explain why I have 7e hair and that I’m in Group 7A on the Periodic Table. The two energy levels explain the meaning of the 2 dots on my body and why I’m in Period 2 on the Periodic Table. The 9 electrons around my nucleus and the 9 protons in my nucleus are what my Atomic Number says I should have. The 7 electrons in the Outside Energy Level explain why I’m in Group 7A. Amazing!”

“Now that you’ve finished drawing my Bohr model, I would like to have you sit down and visit with me and listen to the story of Fluorine. That’s me.”

Professor Terry and Guy are happy to learn about Fluorine. Fluorine moved several comfortable chairs together on the roof by the peach tree. They each enjoyed eating a peach as Fluorine began his story. Guy was happy to know more about another part of his magnificent world.

FLUORINE

Fluorine said, “Most of the products I’m used for are compounds made from the ore fluorite. I’m used as a refrigerant, in electrical insulation and in cookware. I think you’ve heard of teflon used to coat pots and pans to prevent sticking. That’s one of my uses. My very corrosive acid, HF, is used to etch pretty designs into glass. You know about dentists using some of my compounds to strengthen teeth and prevent cavities. For that same reason, some of my compounds are put into toothpaste.”

Guy thanked Fluorine, and he and the professor were off to see Neon. All Guy could think about were those Fluorine compounds building a protective wall around his teeth. It was so good to know what Fluorine’s atom looked like. Fluorine said, “I love my Bohr model.”

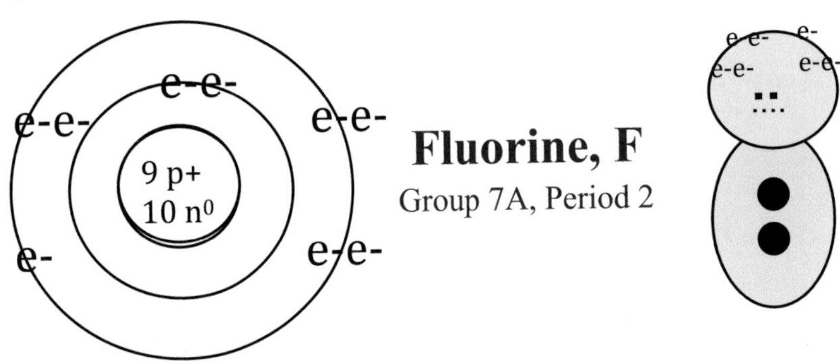

NEON

They went back through the Pine forest. Guy had fun kicking pine cones and enjoying the aroma of pine They finally reached the spot where they had put their bubble and balloons. The ropes that connected the balloons to the bubble were tangled. It took forever to untangle them. When finished, they set off to see Neon.

On arriving at 8th Street, they popped out of the bubble. This time they were more careful with the balloon's ropes. They hung them up carefully over a low branch of a tree. Deciding the ropes were tied up properly, they started looking for Neon's house. His address was #10, 8th Street. "Wait," said Professor Terry, "I think we've forgotten something."

Guy said, "We have to color Fluorine on the map.What's his family color?" Professor Terry said, "It's yellow green."

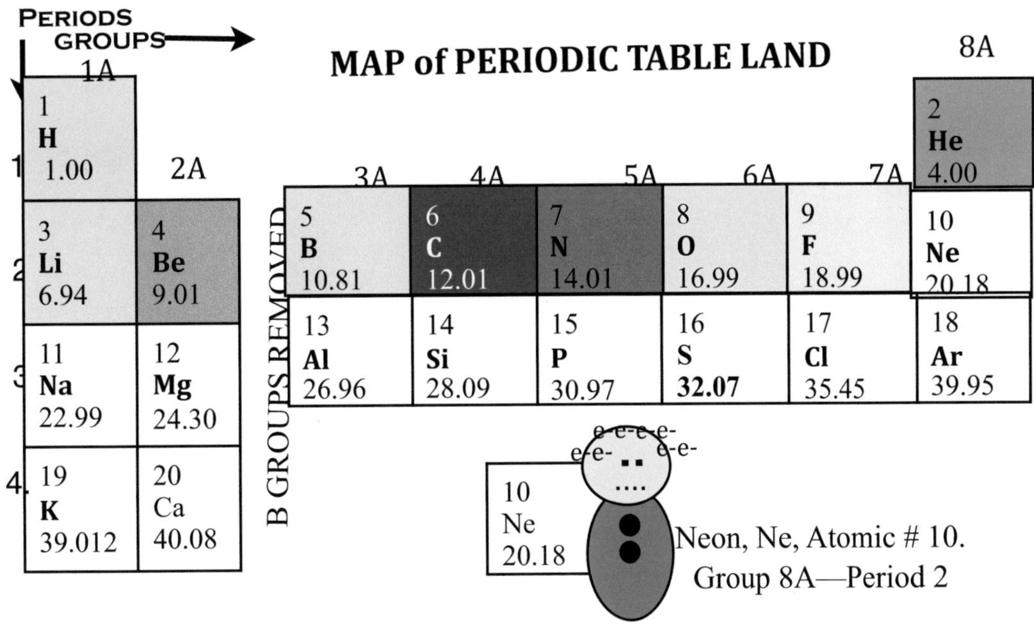

Guy swiftly colored Fluorine's element box yellow green. Then, they were off to Neon's house. There was no doubt that they had the right place. Neon signs hung everywhere. When Neon opened the door, Guy walked up the wide wooden porch steps and said, "We've come to teach you how to draw your Bohr model. As I look at you I see there are some things about the way you look that help me know something about you."

Neon said, "I guess you are talking about my 8e hair and the 2 dots on my body,"

Guy said, "That's right. Your Bohr model will show whether those marks are accurate. It will also confirm the meaning of your Element Box. Then, Guy went on to explain more,"I made a chart for you to keep. It tells step by step how to draw a Bohr model. You should read this while I draw your Bohr model. Following along will help you learn to draw your own Bohr model. Here it is." Then Guy went over to the chalk board and began drawing Neon's Bohr model.

Guy Gets Busy Drawing Neon's Bohr Model

1. Label the Columns

ELEMENT BOX CALCULATIONS BOHR MODEL

2. Copy and Label the Element Box from the Periodic Table.

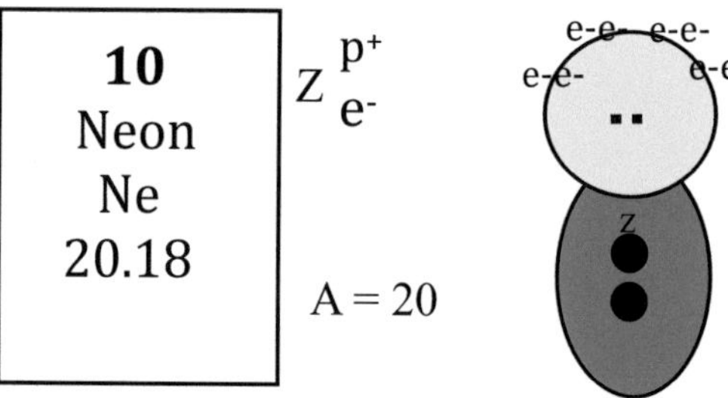

3. Do the Calculations

$n^0 = A - Z$

$n^0 = 20 - 10$

$n^0 = 10$

4. Draw Neon's Atom

1st draw a thick circle for the nucleus.

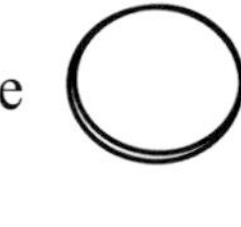

Neon's Atom

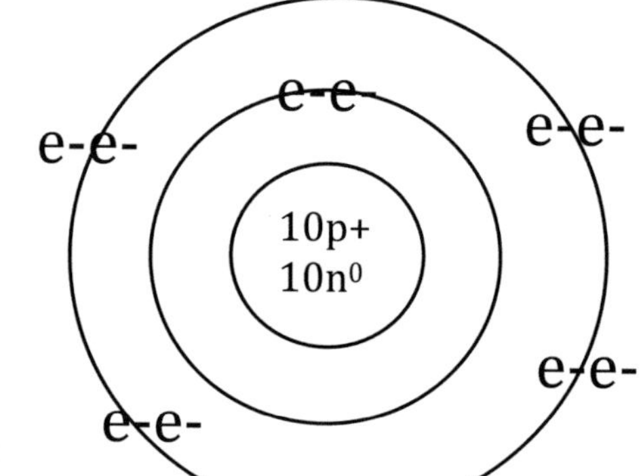

2nd Neon is in **Period 2.**
So, Draw 2 Energy Levels around the nucleus.

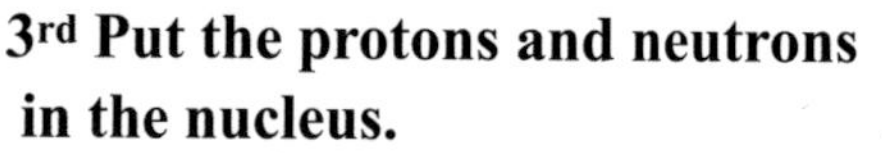

3rd Put the protons and neutrons in the nucleus.
p+ = Atomic # Neon's is 10
n^0neutrons (See Calculations).

10p+
10 n^0

4th Put electrons in Energy Levels.
# of Electrons equals the Atomic Number.
Neon's Atomic Number is 10.
Put 10 electrons in the energy levels.

Guy said, "I've finished drawing your Bohr model. Come look at it."

NEON'S BOHR MODEL

ELEMENT BOX CALCULATIONS BOHR MODEL

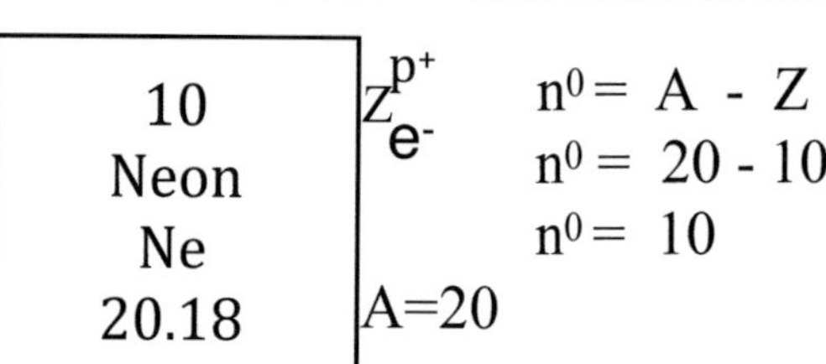

$n^0 = A - Z$

$n^0 = 20 - 10$

$n^0 = 10$

e-e-e-e-
e-e- e-e-
e-e-
e-e-
e-e-
10p+
$10n^0$
e-e-
e-e-

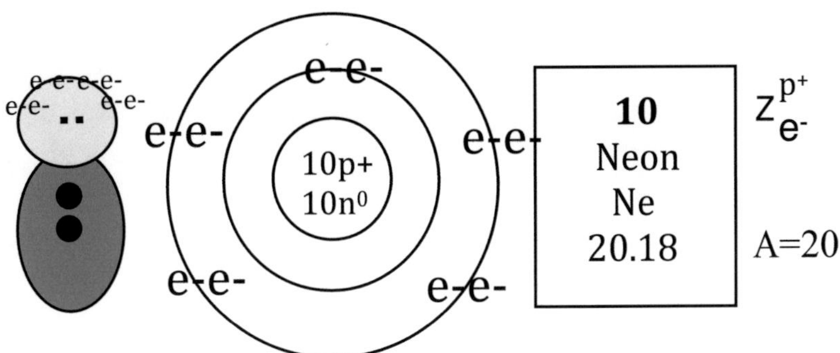

Neon was delighted to see how his Bohr model explained the marks on his body and his Element Box on the Periodic Table. "Check for yourself. My 8e hair represents the 8 electrons in the Outside Energy Level of my atom. That means also I'm in Group 8A, and I am. The 2 dots on my body say I have 2 energy levels, and I am in Period 2. I can actually see the 2 energy levels in my atom. It's all good. My Atomic Number 10 equals the number of protons in the nucleus and the number of electrons in my energy level. I can count them. Yes, the Bohr model tells me so much.

"Now that I have my Bohr model, I would like to have you sit down and visit with me. I'd like to tell you a little about Neon. That's me."

NEON

I am a colorless and odorless gas. I am only 0.0018% of the earth's atmosphere. Although there is very little of me on earth, I am financially valuable. They use me in neon signs. New York City is famous for its colorful Neon signs. I was discovered in 1898 when Ramsay and Travers, two English scientists, were experimenting. When they put me into a vacuum tube and applied high voltage electricity the whole tube filled with a blaze of crimson light. This is the moment in time when I became famous for making beautiful, colored advertising signs. Another one of my uses is to cause helium-neon lasers to look red. Neon is used as a cryogenic. I freeze things to very low temperatures. My element is essentially inert. That means that I am not known to form compounds."

So, this is what my atom looks like!

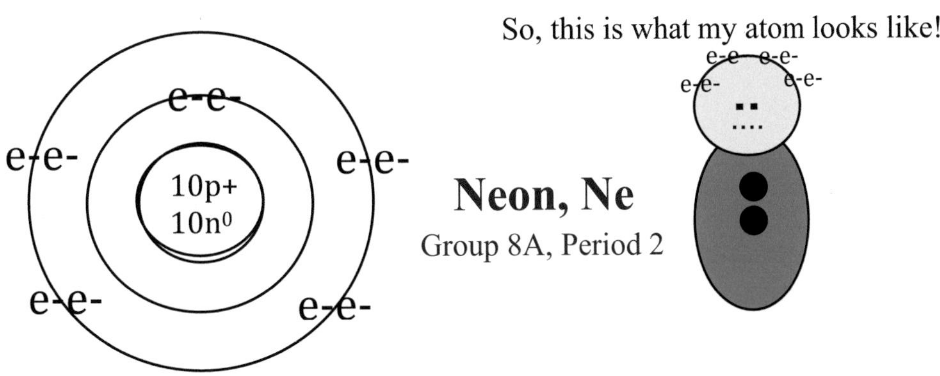

Neon, Ne
Group 8A, Period 2

SODIUM

The scenery was captivating as they flew back to 1st Street to see Sodium. Flying across Periodic Table Land Guy noticed so many lakes and tiny streams that he had never noticed before. He saw many elements enjoying a day in their boats and others fishing from piers. The trip went by fast, and they were fast approaching 1st Street. When they landed, he remembered to color Neon's box purple. showing they had drawn his Bohr model

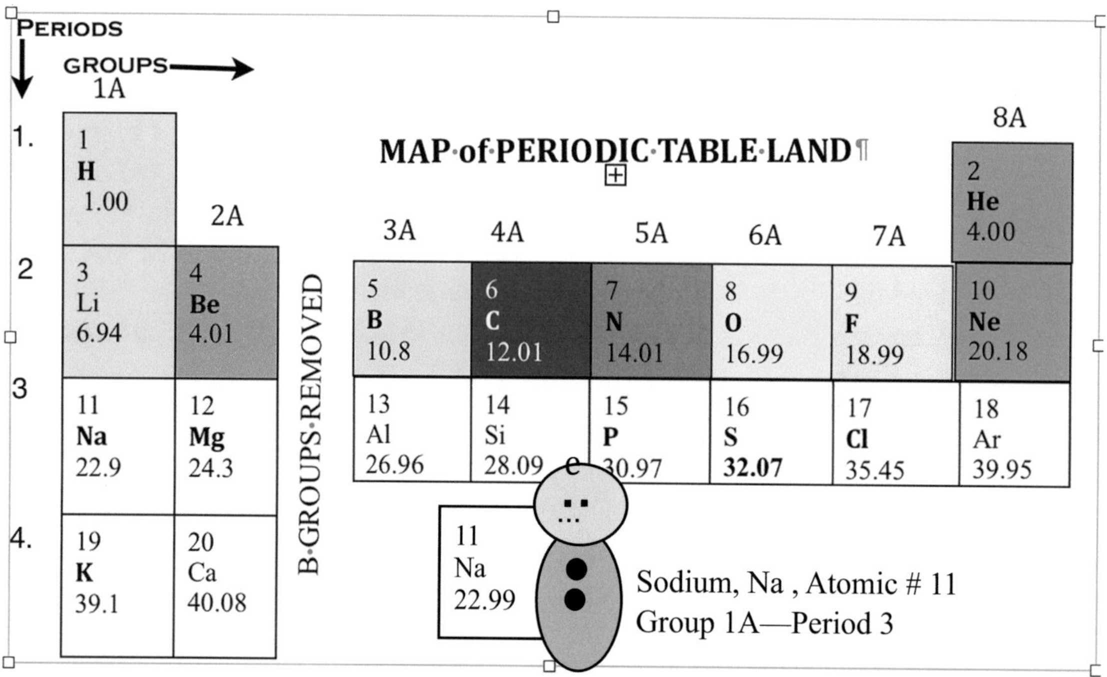

Then they were off to Sodium's house at #11, 1st Street. It was a familiar route as they had walked to Sodium's house often. Passing Lithium's house, they looked ahead and saw their dear friend Sodium standing on his front porch waving to them.

One glance revealed to Guy that Sodium would be so happy to see his Bohr model. It would confirm that his 1e hair meant he had 1 electron in his Outside Energy Level. His triangle would also be confirmed to mean he had 3 energy levels. Professor Terry was right when she predicted our next elements would be in Period 3. However, Guy would be sure when he could actually see 3 energy levels in Sodium's atom.

Sodium was happy to see Guy and Professor Terry again and said, "I heard you've come to draw my Bohr model. Lithium and Hydrogen both told me you had done theirs." Guy said, "When you see your Bohr model you will understand better the information in your Element Box on the Periodic Table and the meaning of your body markings." Guy showed Sodium the Steps Chart he had created to describe how a Bohr model is made. Guy explained, "This chart will make it easier for you to follow along as I draw your Bohr model. Step by step, you will learn how to draw your own Bohr model. Here's the chart. Let's hang it close to where I'll be drawing your Bohr model." Then Guy began.

Guy Gets Busy Drawing Sodium's Bohr Model

1. Label the Columns:

ELEMENT BOX CALCULATIONS BOHR MODEL

2. Copy and Label the Element Box from the Periodic Table.

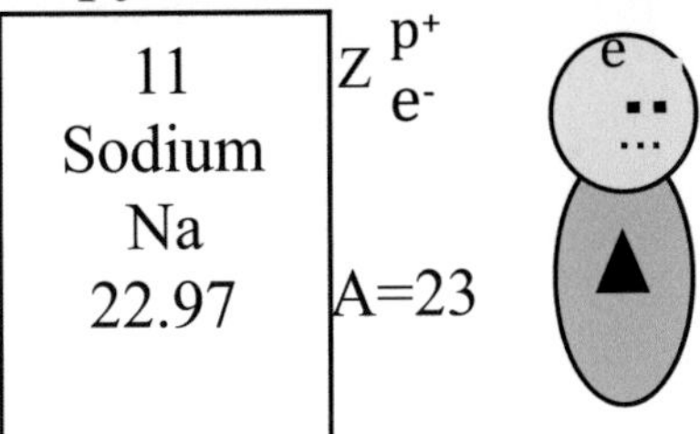

3. Do the Calculations

$n^0 = A - Z$

$n^0 = 23 - 11$

$n^0 = 12$

4. Draw Sodium's Atom

1st draw a thick circle for the nucleus.

2nd Sodium is in **Period 3**, **So, Draw** 3 Energy Levels around the nucleus.

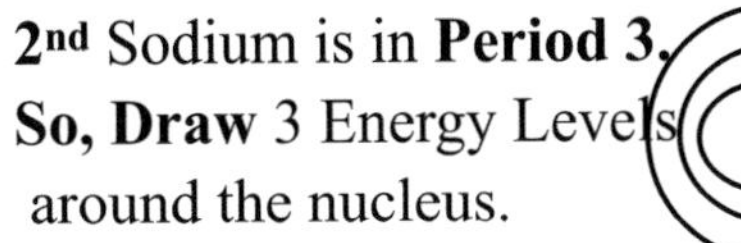

Sodium's Atom

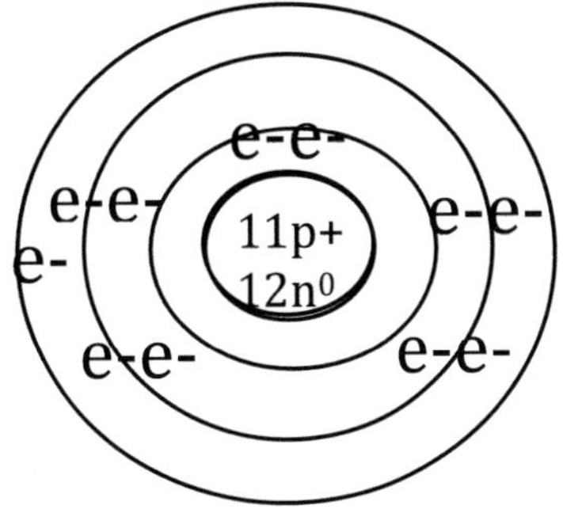

3rd Put the protons and neutrons in the nucleus.

p+ = Atomic # Sodium's is 11 11p+

n^0neutrons (See Calculations. $12n^0$

4th Put electrons in Energy Levels.

# of Electrons equals the Atomic Number.

Sodium's Atomic Number is 11.

Put 11 electrons in the energy levels.

Guy said, "I've finished drawing your Bohr model. Come look at it."

SODIUM'S BOHR MODEL

ELEMENT BOX CALCULATIONS\ BOHR MODEL

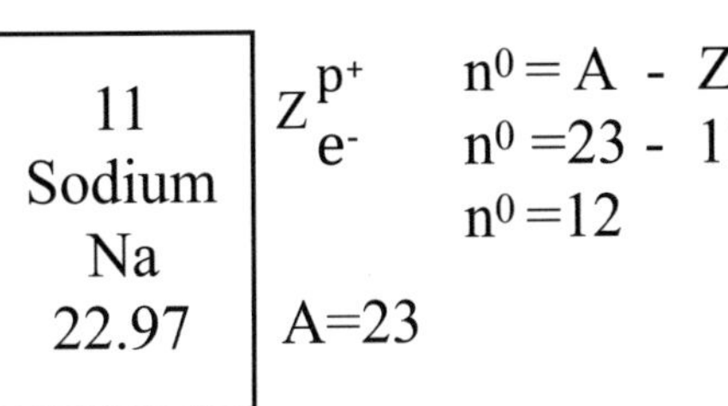

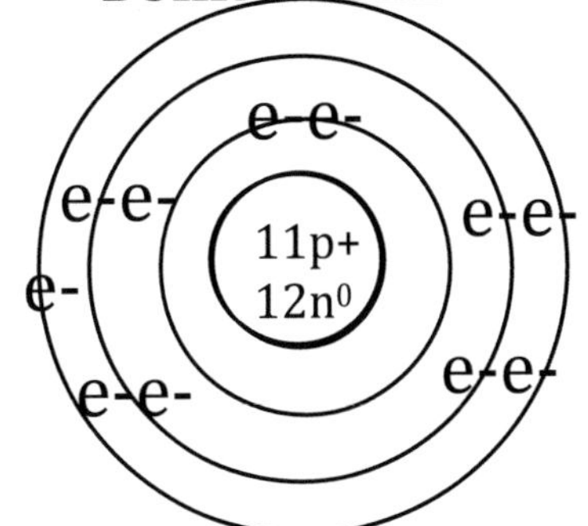

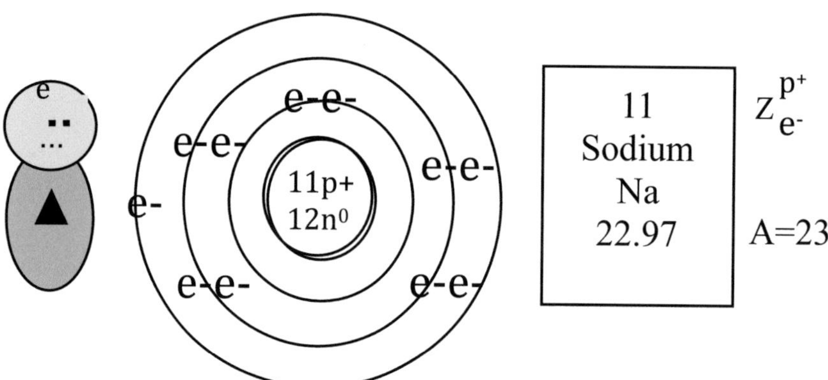

Sodium checked out his newly drawn Bohr model to see what it was all about. He saw that the Atomic Number, 11 did tell he had 11 protons in the nucleus and 11 electrons in the energy levels. My 1e hair is that 1 electron in the Outside Energy Level. It also means I'm in Group 1A. The 3 energy levels confirm the fact that I'm in Period 3. That's why I have the triangle on my body. I love my Bohr model. It tells me so much." With his Bohr model understood, Sodium invited Professor Terry and Guy to come into his house and learn some more information about his element, his compounds and his uses.

SODIUM

Sodium announced, "I'm a soft, silvery metal. I'm the 6th most abundant element in the earth's crust. I'm mainly found in the compound salt, Sodium chloride. Since my salt dissolves easily in water, I'm found in earth's oceans helping to make them salty. I was discovered by Sir Humphrey Davy in 1807. I'm in the compound, sodium hydroxide. That is lye, a product used to make soap. I am an essential element in living things: people, animals and some plants. In our bodies I affect blood pressure. If you have high blood pressure, doctors sometimes tell you to use less salt. I am found in Baking Soda (Sodium bicarbonate) that is used to make your stomach less acid. It's also used to absorb odors in the refrigerator. Mothers keep handy a tiny jar of sodium bicarbonate dissolved in water to remove the odor of baby spit up on their clothes." Off they went to see the next element Atomic Number,12.

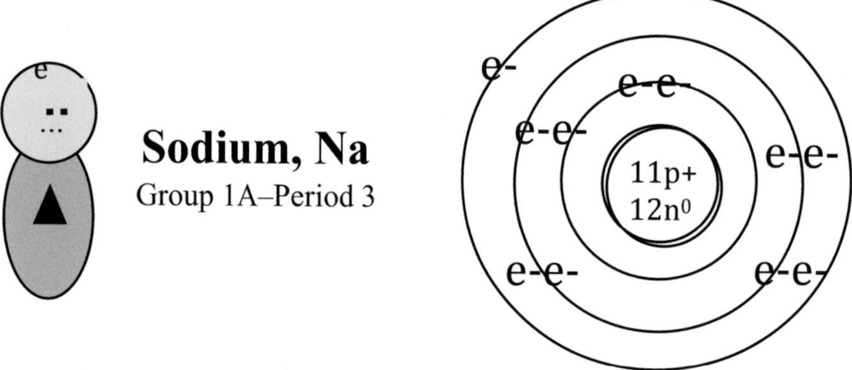

MAGNESIUM

It was a beautiful summer day. Flying around in the bubble was amazingly relaxing. Guy looked at the clouds, the mountains in the distance and the colorful flowers surrounding the houses below. Guy shared with Professor Terry how happy he was to be learning chemistry. "It's especially wonderful to be looking at the models of the atoms that make up my beautiful world. I'm learning so much."

Professor Terry reminded Guy to color in Sodium's box on the map since they just finished his Bohr model. Guy could see that Sodium was in Group 1A with Hydrogen and Lithium. It was obvious that green was their family color.

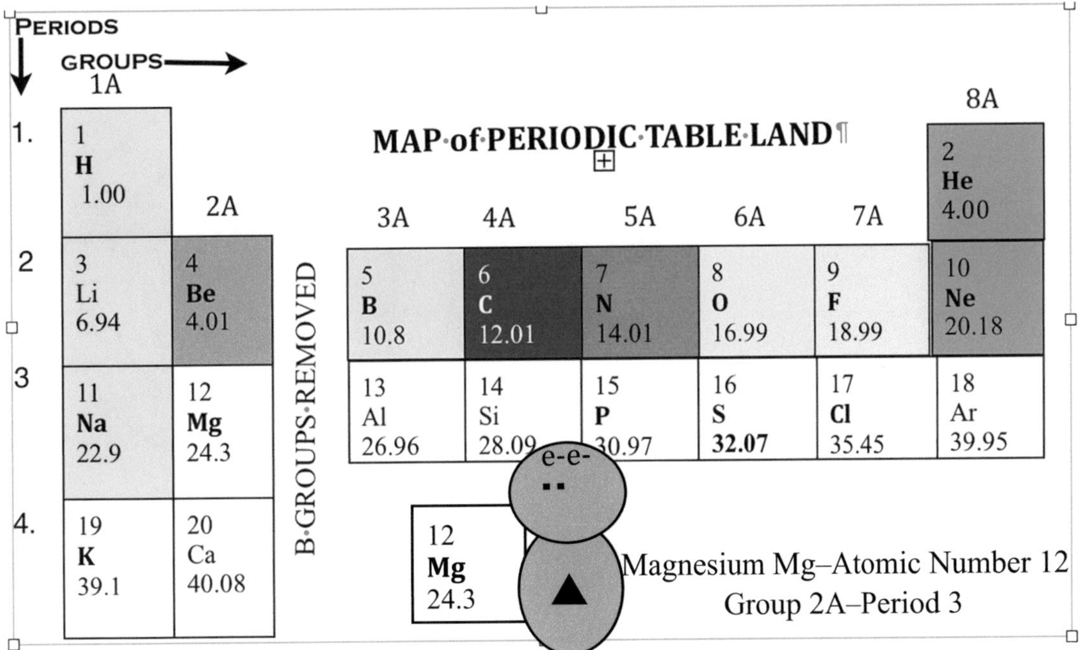

They landed, and off they went to see Magnesium at #12, 2nd Street. Magnesium opened the door as soon as they knocked. He was an adorable little brown element. Brown was the family color on 2nd Street. He had a triangle on his body. Remembering that a triangle has 3 sides, Guy knew that Magnesium had 3 energy levels. His 2e hair showed he was in Group 2A. Guy pointed out to Magnesium that when his Bohr model was drawn it would verify the meaning of these body markings and help him understand better his Element Box on the Periodic Table.

Professor Terry introduced Guy to Magnesium since this was the first time they were meeting. Then, Guy explained that he had come to draw his Bohr model. Guy showed him the Steps Chart he had created explaining how it would help him learn to make a Bohr model himself. Magnesium had a large drawing pad laid out on his wide flat table. Then he produced a sharpie for Guy to use to draw his Bohr model. The Step Chart hung, Magnesium dragged over a stool and placed it where he could see both the Steps Chart and the pad on which Guy would be drawing the Bohr model. Then Guy began.

Guy Starts to Draw Magnesium's Bohr Model

1. Label the Columns:

ELEMENT BOX CALCULATIONS BOHR MODEL

2. Copy and Label the Element Box from the Periodic Table.

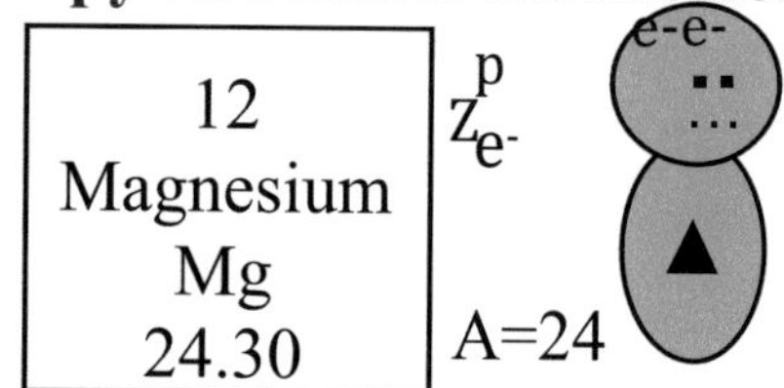

3. Do the Calculations

$n^0 = A - Z$

$n^0 = 24 - 12$

$n^0 = 12$

4. Draw Magnesium's Atom

1st draw a thick circle for the nucleus.

2nd Magnesium is in **Period 3.** **So, Draw** 3 Energy Levels around the nucleus.

3rd Put the protons and neutrons in the nucleus.

p+ = Atomic # Magnesium's is 12 — 12 p+

n^0neutrons (See Calculations). — 12 n^0

4th Put electrons in Energy Levels.

# of Electrons equals the Atomic Number.

Magnesium's Atomic Number is 12.

Put 12 electrons in the energy levels.

Magnesium's Atom

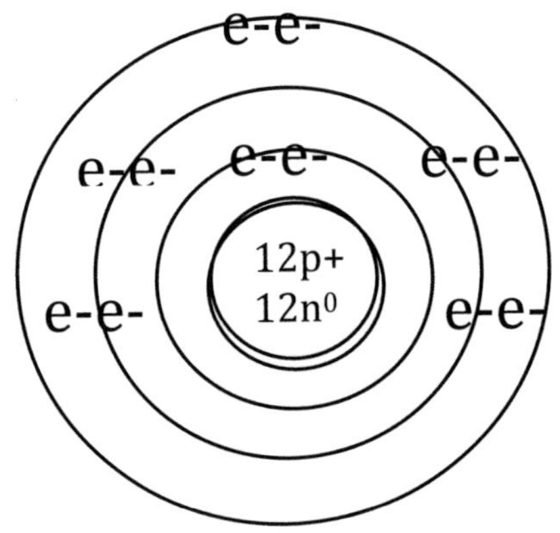

Guy said, "I've finished drawing your Bohr model. Come look."

MAGNESIUM'S BOHR MODEL

ELEMENT BOX **CALCULATIONS** **BOHR MODEL**

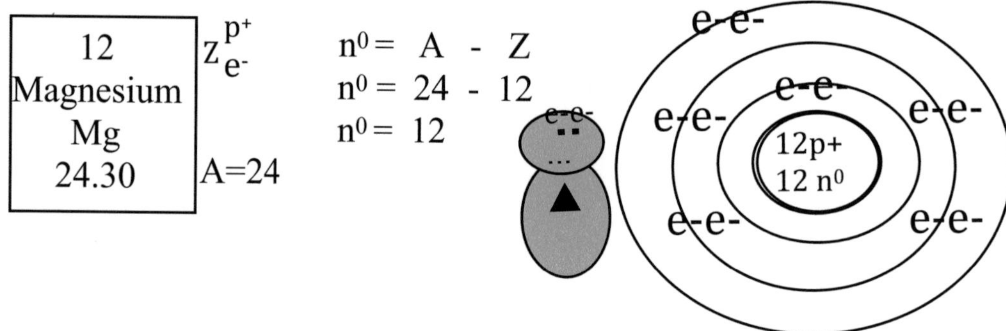

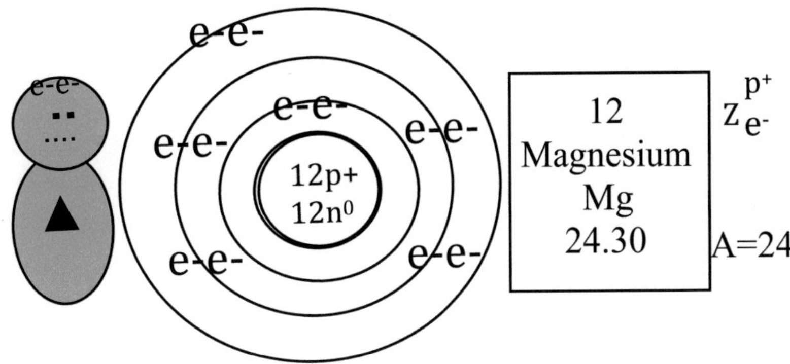

Magnesium was so happy. Looking at his Bohr model he observed, "12 electrons in my energy levels, and 12 protons in my nucleus. It's just as my Atomic Number 12 in my Element Box predicted. Wonderful!"

Guy said, "Look carefully at the 2 electrons in your atom's Outside Energy Level. They mean you are in Group 2A on the Periodic Table. Your 2e hair also represents those same 2 electrons in your Outside Energy Level."

Magnesium added, "I see that I do have 3 energy levels in my atom. That shows I really am in Period 3. That's what the triangle on my body announces too. Now I know it's correct." Magnesium was thrilled that his Bohr model showed so much.

MAGNESIUM

"Now that I understand my Bohr model, I would like to have you sit down and visit with me. I'd like to tell you a little about my element, some of the compounds that my element is a part of and my uses." So they followed Magnesium into his house, sat down and listened.

Magnesium, Atomic number 12 said,"I make the brightest white fireworks. I'm lots of fun on the 4th of July. In early days of photography, flashbulbs used magnesium to produce a bright light to take pictures. Magnesium powder is used in marine flares and to create theatrical effects. Magnesium is an important element in helping the body to function properly. Two of my compounds that make useful products are Epsom salts and Milk of Magnesia." Guy enjoyed learning about Magnesium. Then they left.

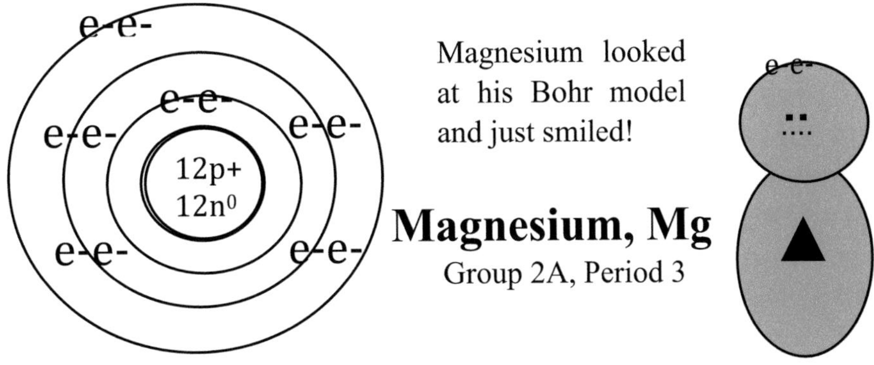

Magnesium looked at his Bohr model and just smiled!

Magnesium, Mg

Group 2A, Period 3

ALUMINUM

It was a long flight over wide B Avenue. Guy looked down on the homes of the true metals that lived there. Guy noted that although they were all metals, it was amazing how different their houses were. Then he thought about the element, Magnesium they had just left. Professor Terry reminded Guy to take out the map and color in Magnesium's box to show that they had finished his Bohr model. Guy could see that Magnesium was in the same group as Beryllium, the element above Magnesium, Group 2A. Professor Terry had told him that all the elements in the same A Group are in the same family, and they all had the same family color. So he colored Magnesium brown like Beryllium. Now, he knew a little more about another element that made up his world.

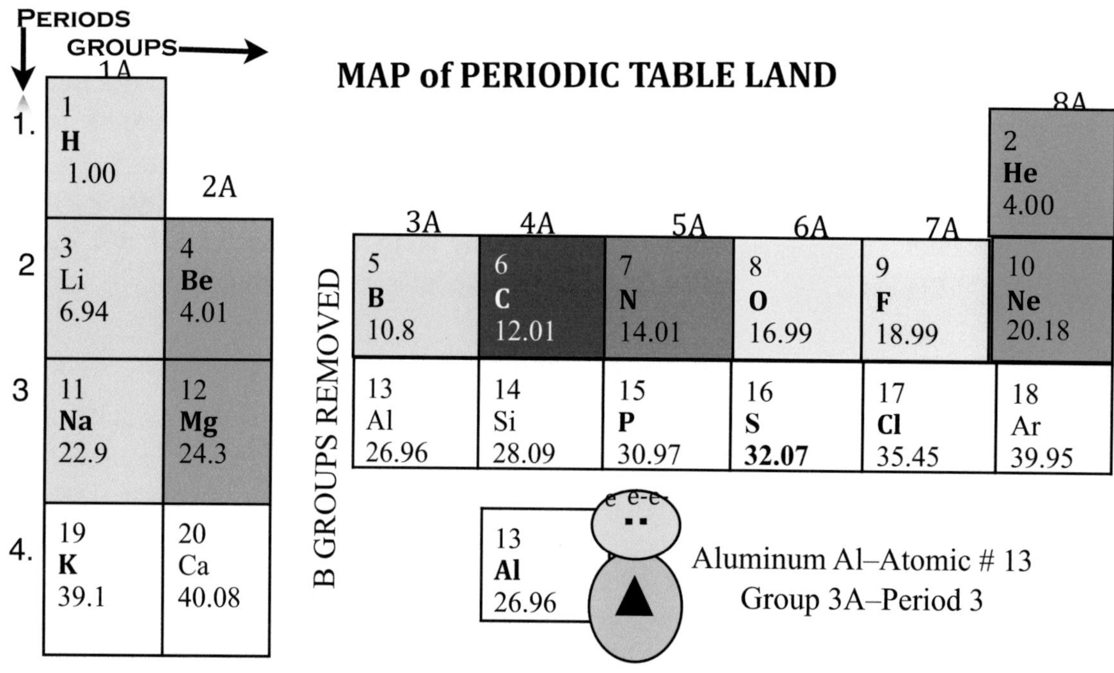

They landed on the tarmac behind 3rd Street, and off they went to see Aluminum at #13, 3rd Street. When Aluminum opened the door soon after they knocked Guy noticed he was gray. Now he knew that was his family color. Professor Terry explained that Guy had come to draw his Bohr model. Guy pointed out to Aluminum that when his Bohr model was drawn it would verify the meaning of his body markings and help him understand his Element Box.

Guy had visited Aluminum a while back, and Aluminum remembered him saying, "I remember you counted my 13 electrons and saw my 13 protons and you were happy to see they equaled my Atomic Number." Guy showed him the Steps Chart he had created to help him learn how to draw a Bohr model by himself. After touring the house Aluminum lead them to his favorite room with soft sunlight streaming in. Here Aluminum had everything ready—a white board, a black marker, and a place to hang the Steps Chart. With everything ready, Guy got started right away.

Guy Gets Busy Drawing Aluminum's Bohr Model

1. Label the Columns:

ELEMENT BOX CALCULATIONS BOHR MODEL

2. Copy and Label the Element Box from the Periodic Table.

13 Aluminum Al 26.15

$Z^{p^+}_{e^-}$

A=26

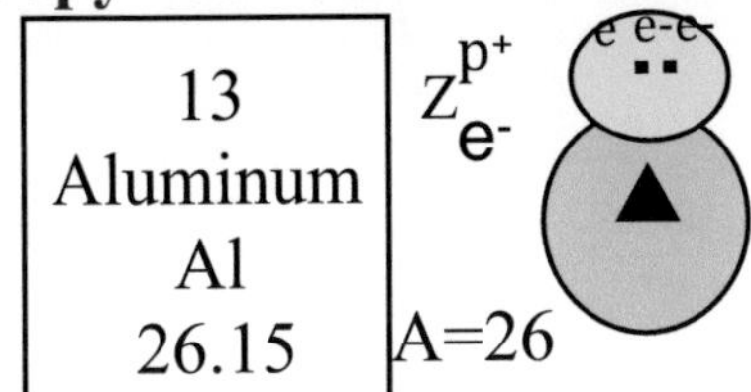

3. Do the Calculations

$n^0 = A - Z$

$n^0 = 26 - 13$

$n^0 = 13$

4. Draw Aluminum's Atom

1st draw a thick circle for the nucleus.

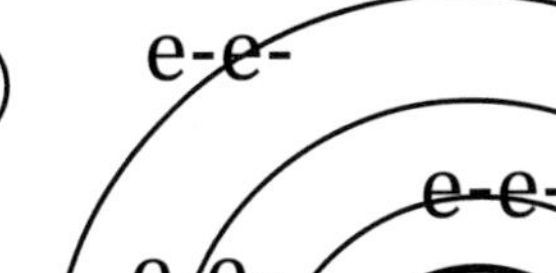

2nd Aluminum is in **Period 3.**
Draw 3 Energy Levels around the nucleus.

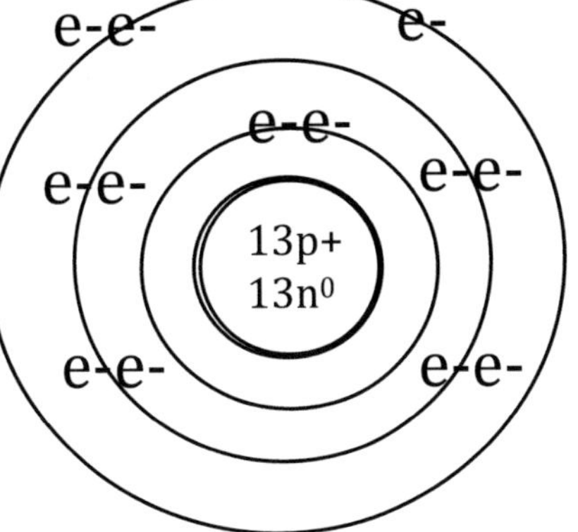

3rd Put the protons and neutrons in the nucleus.

p+ = Atomic # Aluminum's is 13 13p+

n^0neutrons (See Calculations). 13n^0

Aluminum's Atom

e-e- e- e-e- e-e- e-e- e-e- e-e- 13p+ 13n^0

4th Put electrons in Energy Levels.

# of Electrons equals the Atomic Number.

Aluminum's Atomic Number is 13.

Put 13 electrons in the energy levels.

Guy said, "I've finished drawing your Bohr model. Come look at it."

ALUMINUM'S BOHR MODEL

ELEMENT BOX CALCULATIONS BOHR MODEL

13 Aluminum Al 26.15

$Z^{p^+}_{e^-}$

A=26

$n^0 = A - Z$

$n^0 = 26 - 13$

$n^0 = 13$

e-e- e- e-e- e-e- e-e- e-e- e-e- 13p+ 13n^0

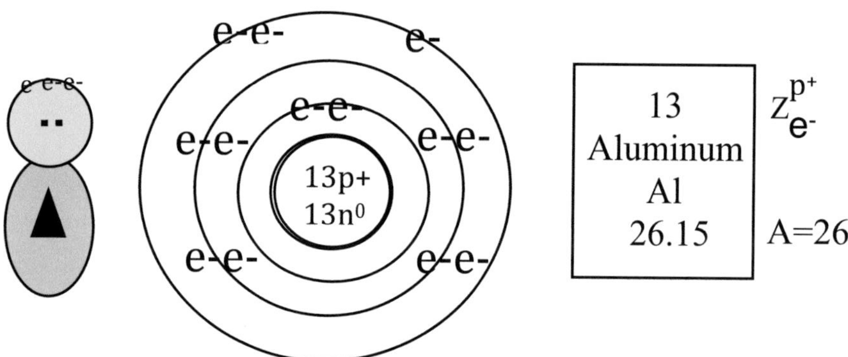

Aluminum said, "I always knew I was in Period 3. That meant having 3 energy levels. However, it's so much better to see the 3 energy levels in my Bohr model. Being in Group 3A, I knew I should have 3 electrons in my Outside Energy Level, but now I actually see those 3 electrons there. My Atomic Number is 13. Now, I see the 13 electrons in the energy levels and 13 protons are in the nucleus of my atom. I can actually see where the electrons are in my atom. Thank you for teaching me how to draw my atom."

Aluminum was delighted to see his Bohr model and said, "Now that you've finished drawing my Bohr model, I would like to have you sit down and visit with me. I'd like to tell you a little bit about myself, my compounds and my uses."

ALUMINUM

Aluminum said, "I'm a silvery white metal. Because I'm light weight, strong and malleable, I can be hammered into sheets of flat metal. I'm used in airplanes, cars, and bicycles. Because I don't corrode I'm used as siding for houses and the weather doesn't affect me. I'm ductile, so I can be pulled into wire. I am conductive. That means, I can carry an electric current through wire. However, I like best being in so many people's kitchens, providing them with utensils because I'm not toxic."

Guy added, "My father uses aluminum foil on the grill. My mom makes cookies on an aluminum cookie sheet. Our family recycles aluminum cans."

When Guy and Professor Terry left, Aluminum never tired of looking at his atom.

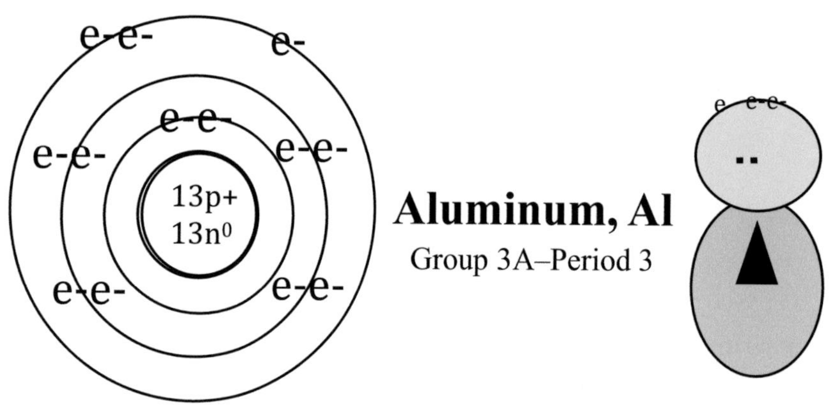

SILICON

Professor Terry said, "Before we leave here, let's color Aluminum's box on our map. Aluminum is in Group 3A so his family color is light gray."

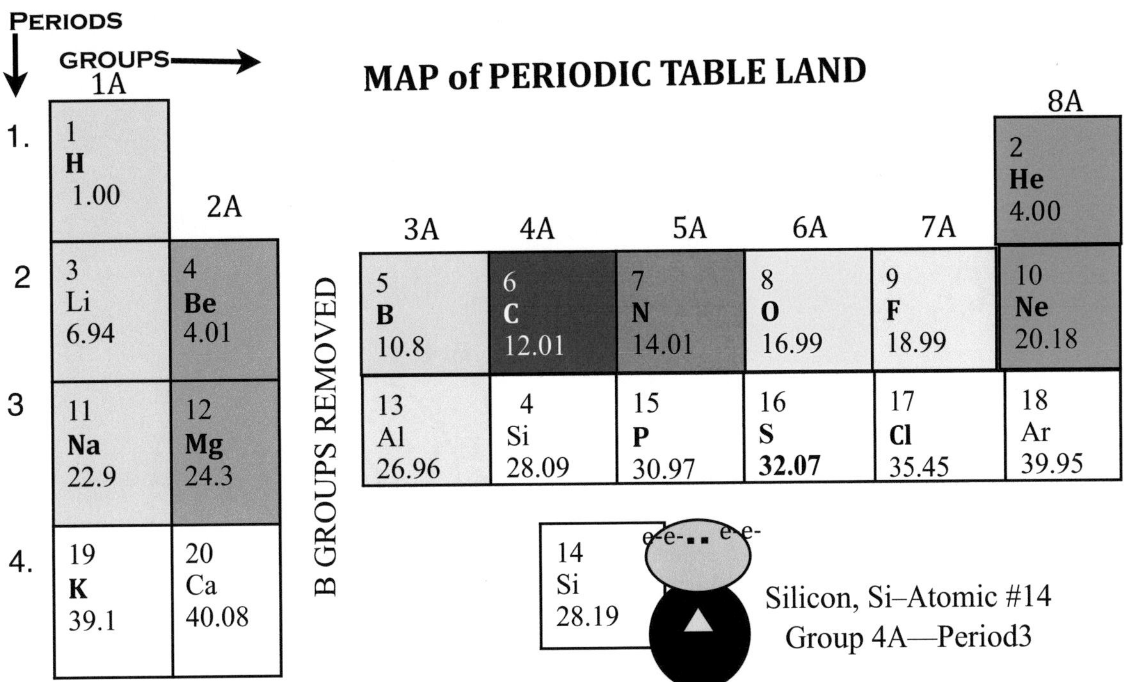

Aluminum's box colored, they flew low over the house tops, and arrived at 4th Street rather fast. Guy said, "I see Silicon's house is surrounded by sand. His house is a sandcastle. Guy hoped to be invited to tour the house. He was so curious. They put away the bubble and balloons safely, and made their way beyond Carbon down to # 14, 4th Street. Guy's dream came true. Silicon gave him a tour of his Sandcastle. Guy was spellbound. Inside the furniture was all molded out of hard sand and sprayed to make it keep its shape. His bed was molded to the shape of his body. It must have been so comfortable.

Silicon said, "I know you came to draw my Bohr model. I heard that's what you've been doing with all the other elements."

Guy responded, "When you get to see the model of your atom, you will understand the meaning of your Element Box on the Periodic Table as well as your body markings so much better."

Silicon said, "That's what everyone's been saying."

Then Guy added, "I guess it's time to draw your Bohr model. I created a chart that will help you learn how to draw your own Bohr model after I'm gone. Let's decide on a place to put this chart so you will be able to see it and follow along while I draw your Bohr model."

Silicon was always organized When they entered his den, there was a white board set up next to the desk where Silicon would sit. They hung the Steps Chart nearby. Guy picked up the marker and began.

Guy Gets Busy Drawing Silicon's Bohr Model

1. Label the Columns:

ELEMENT BOX CALCULATIONS BOHR MODEL

2. Copy and Label the Element Box from the Periodic Table.

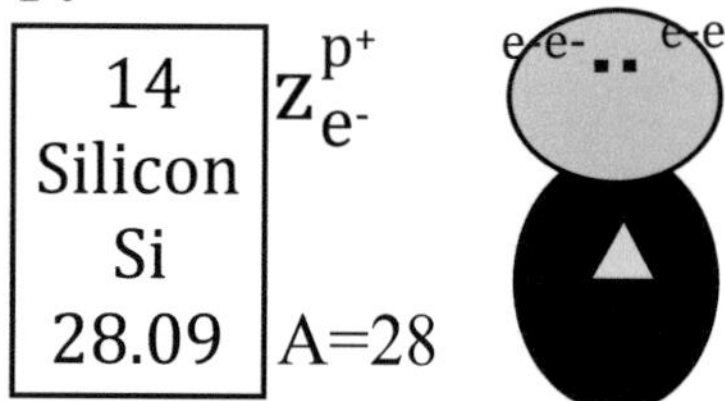

3. Do the Calculations

$n^0 = A - Z$

$n^0 = 28 - 14$

$n^0 = 14$

4. Draw Silicon's Atom

1st draw a thick circle for the nucleus.

2nd Silicon is in **Period 3.** **So, Draw** 3 Energy Levels around the nucleus.

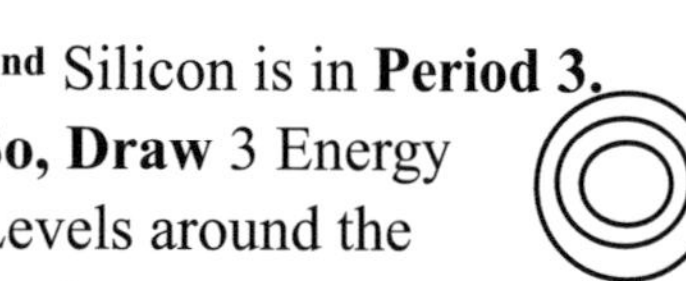

3rd Put the protons and neutrons in the nucleus.

p+ = Atomic # Silicon's is 14 14p+

n^0neutrons (See Calculations). $14n^0$

Silicon's Atom

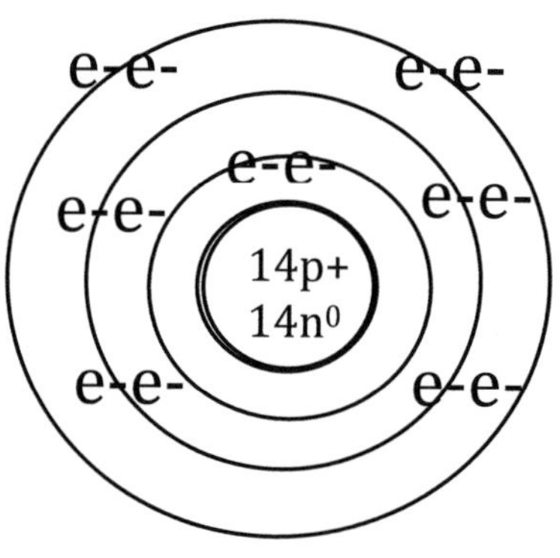

4th Put electrons in Energy Levels.

# of Electrons equals the Atomic #.

Silicon's Atomic Number is 14.

Put 14 electrons in the energy levels.

Guy said, "I've finished drawing your Bohr model.

Come look at it."

SILICON'S BOHR MODEL

ELEMENT BOX

14 Silicon Si 28.09	$Z^{p^+}_{e^-}$ A=28

CALCULATIONS

$n^0 = A - Z$

$n^0 = 28 - 14$

$n^0 = 14$

BOHR MODEL

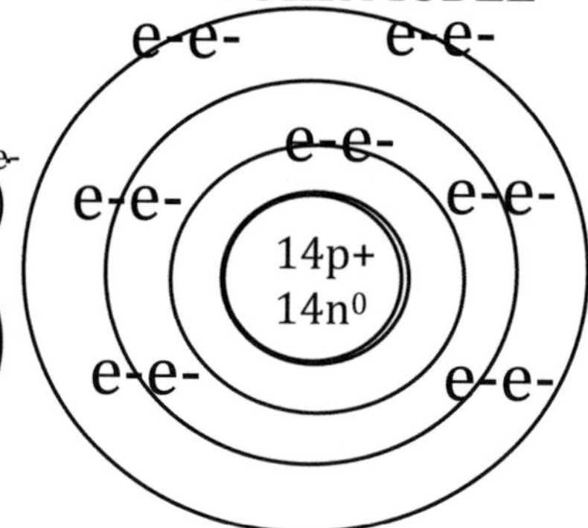

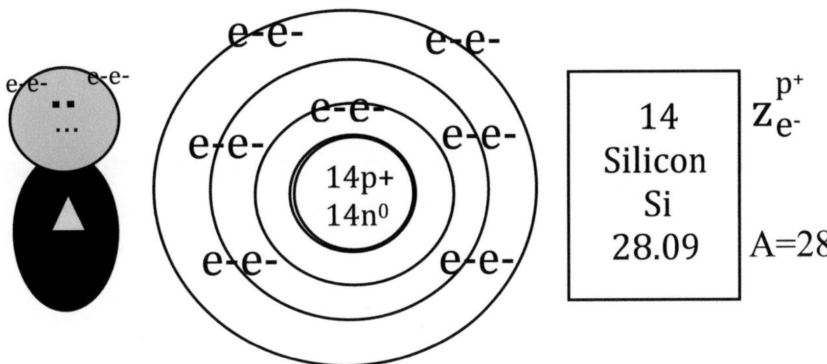

Silicon looked at his Bohr model. He counted the energy levels and indeed there were 3 as the triangle on his body indicated. This proved also that he was in Period 3 on the Periodic Table. Silicon counted the 4 electrons in the Outside Energy Level of his atom and said, "I've always been told I was in Group 4A. These 4 electrons in the Outside Energy Level are what Group 4A means. It's also why I have 4e hair. My Atomic Number is 14. Checking, I do have 14 protons in my nucleus and 14 electrons in my energy levels. It's wonderful how my Bohr model confirms what I always knew."

Guy was glad Silicon was able to look at his Bohr model and see all the information that it offered. Then he said, "I love your home. I've built many sand castles at the beach. Now, I know I've had all this fun thanks to you, Silicon.

Silicon said, "Thank Oxygen, too. Together we make sand in the compound, Silicon dioxide."

SILICON

Silicon said, "Let's sit down and visit a while." Silicon began, "I have some very different kinds of uses. Like I told you, I'm in the compound Silicon dioxide which is sand. I'm used to make glass and pottery. I'm found in cement which is used to pave sidewalks and driveways. When I am used in semiconductors, I need to be absolutely pure. I'm used to make computer chips. Small amounts of me make your bones strong along with Calcium."

Silicon stood there and admired his Bohr model after Guy and Professor Terry had left. He thought, "This little Bohr model contains a lot of information about me."

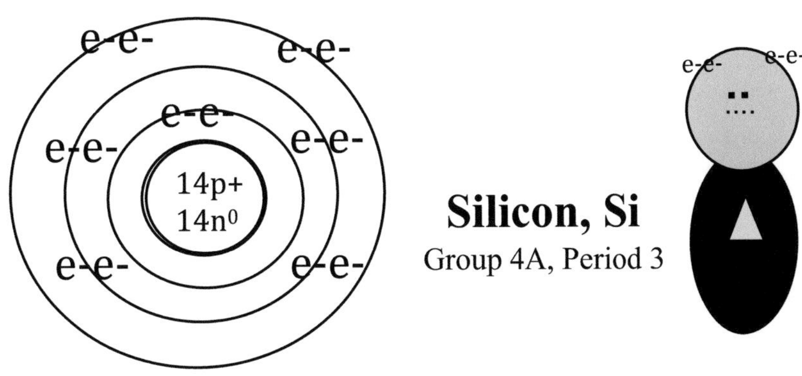

PHOSPHORUS

As Professor Terry and Guy left Silicon, Guy remembered to color in Silicon's box. He saw that Silicon was in the same Group as Carbon, so he knew Silicon's family color was black. The box colored, they jumped into the magic bubble and off they went. They were headed toward 5th Street to see Phosphorus.

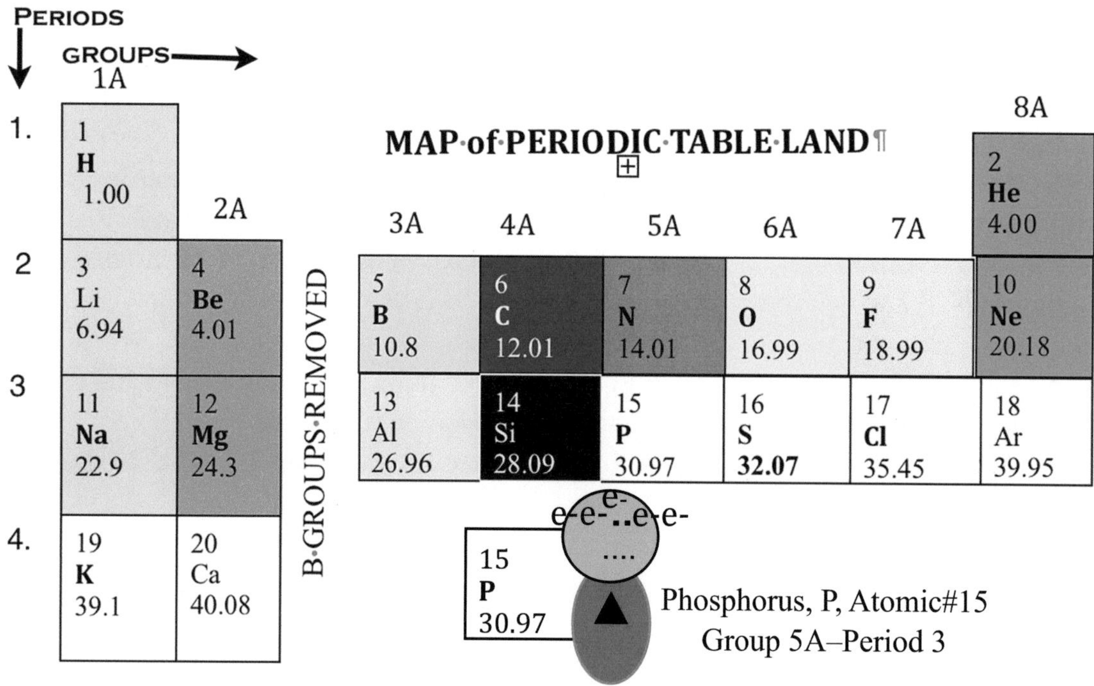

As they approached 5th Street there were so many colorful flowers. Phosphorus as well as Nitrogen had compounds used as fertilizers. No wonder their gardens were so beautiful. They were flying low so the sweet aroma of roses surrounded them with a happy smell. They landed and made their way to #15, 5th Street, the home of Phosphorus.

When they opened the gate to the garden, there was Phosphorus planting some pretty petunias to line the edge of the brick pathway. Guy was quick to observe the 5 electrons he had for hair. This meant that he was in Group 5A and the triangle on his body announced that he was in Period 3. Guy told Phosphorus "When you see your Bohr model, you will learn so much about your body markings and your Element Box on the Periodic Table." Professor Terry explained that Guy would be drawing his Bohr model Phosphorus was not surprised because word spread that Guy was doing this. Phosphorus was eager to see what his atom looked like.

Guy showed Phosphorus the Steps Chart he had drawn to teach the elements how to draw their own Bohr models. Guy said, "I saw a perfect place in your garden to draw your Bohr model and have you check out this chart as I do it." Surrounded by sweet smelling flowers, he hung the Step Chart and began drawing the Bohr model for Phosphorus.

Guy Draws Phosphorus' Bohr Model

1.Label the Columns:

ELEMENT BOX CALCULATIONS BOHR MODEL

2. Copy and Label the Element Box from the Periodic Table.

15 Phosphorus P 30.97

Z p^+ e^-

A=31

3. Do the Calculations

$n^0 = A - Z$

$n^0 = 31 - 15$

$n^0 = 16$

4. Draw Phosphorus' Atom

1st draw a thick circle for the nucleus.

Phosphorus Atom

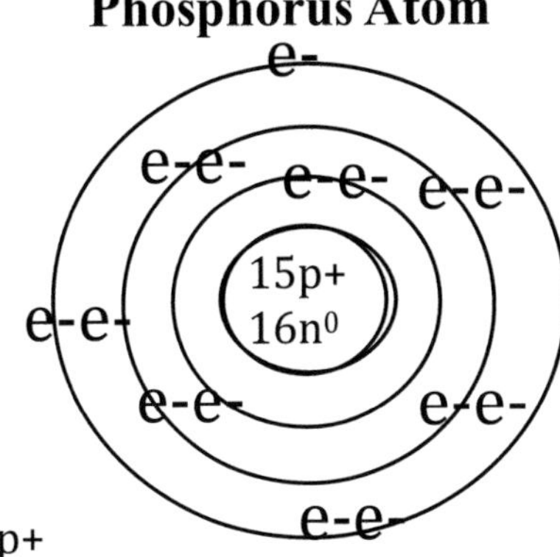

2nd Phosphorus is in **Period 3.**
So, Draw 3 Energy Levels around the nucleus.

3rd Put the protons and neutrons in the nucleus.

p+ = Atomic # Phosphorus' is 15 15p+

n^0neutrons (See Calculations). $16n^0$

4th Put electrons in Energy Levels.

\# of Electrons equal the Atomic Number.
Phosphorus' Atomic Number is 15.
Put 15 electrons in the energy levels.

Guy said, "I've finished drawing your Bohr model. Come look at it."

PHOSPHORUS' BOHR MODEL

ELEMENT BOX CALCULATIONS BOHR MODEL

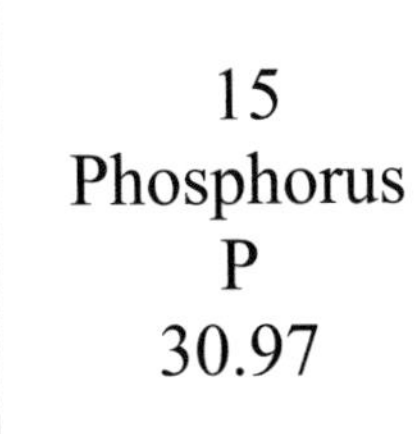

Z p^+ e^-

A=31

$n^0 = A - Z$

$n^0 = 31 - 15$

$n^0 = 16$

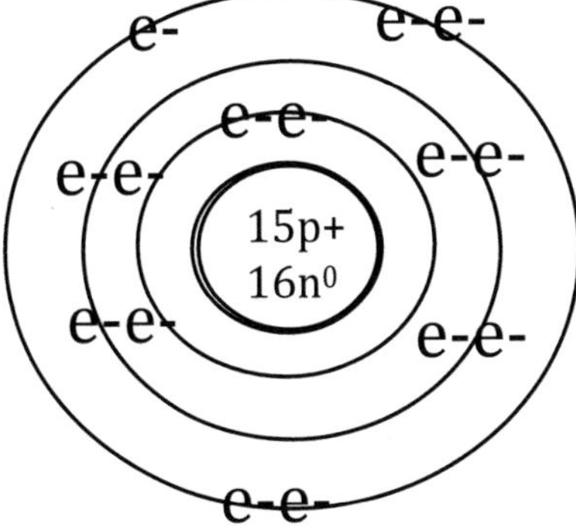

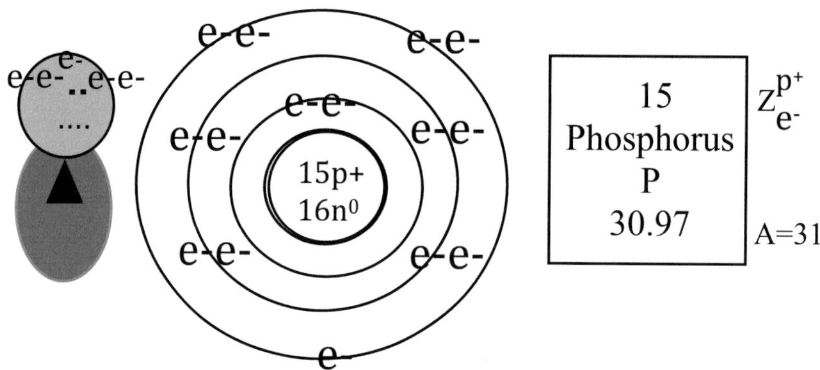

Phosphorus looked at his Bohr model and saw the 15 protons in the nucleus. No other element in the whole universe has 15 protons in their nucleus. "That's what my Atomic Number means. I'm Phosphorus." He also noted there were 15 electrons in the energy levels. As if understanding it for the first time he exclaimed, "That's why I need 3 energy levels to fit in all my electrons. That's what Period 3 means. It's so exciting to see I really have 3 energy levels. My body triangle says I have 3 energy levels. My atom shows this is true. There are 5 electrons in my atom's Outside Energy Level. This explains why I have 5e hair, and that I'm in Group 5A on the Periodic Table."

Professor Terry told Guy, " I want you to pay special attention to the 5 electrons in Phosphorus' Outside Energy Level. I point these electrons out because later when you are learning about how compounds are formed these electrons will play an important role.

PHOSPHORUS

Phosphorus said, "I'm an interesting element. I exist in two forms, white and red. **White Phosphorus** is known for the fact that it glows. This glowing effect is known as chemiluminescence. There are some jellyfish seen glowing in the ocean because of the presence of Phosphorus. I'm also found in glow worms that glow in dark caves. The military uses white Phosphorus for smoke screens. **Red Phosphorus** is noted for its use in matches. Compounds of Phosphorus are found in DNA, in bones, and detergents. My compounds are also used as fertilizers. I have quite a variety of uses."

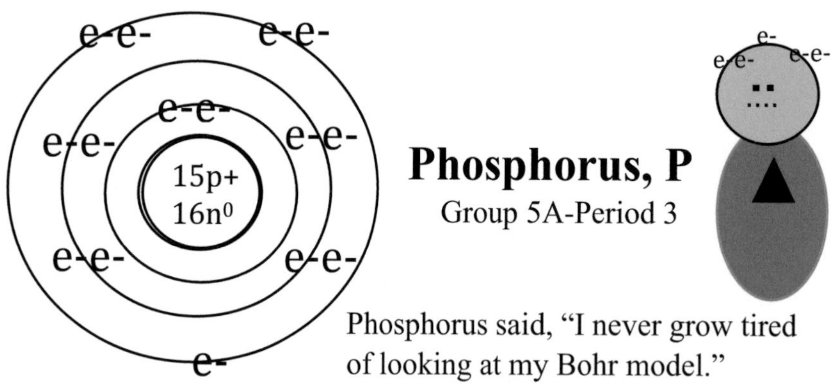

Phosphorus, P

Group 5A-Period 3

Phosphorus said, "I never grow tired of looking at my Bohr model."

SULFUR

Before they got on their way to see Sulfur, Guy remarked, "Of all the elements we visited, Phosphorus is the most interesting. I'd love to see a glow worm!"

Professor Terry agreed and said, "Now let's get Phosphorus' box colored. Notice that Phosphorus is in Group 5A like Nitrogen. So the family color is red. Now it's time to get into the bubble and fly over to 6th Street. I heard Sulfur is expecting us."

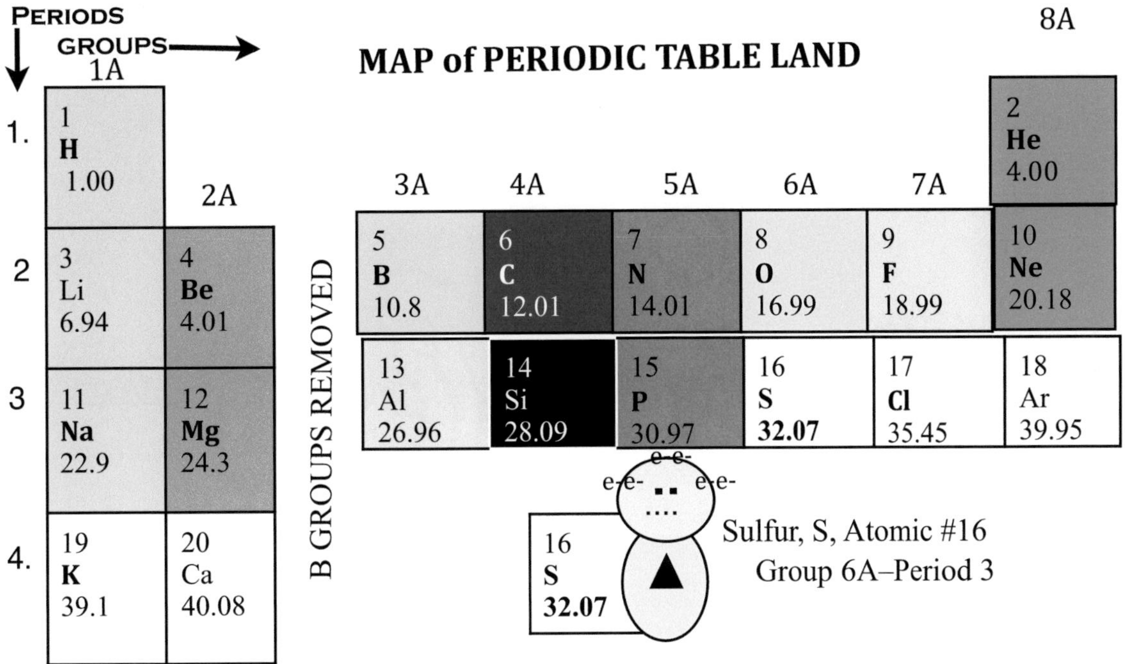

As the balloons lifted the bubble high in the sky, Guy was so happy. The air was fresh, the sun not too warm, it was a beautiful summer day. He gazed over toward 6th Street. Surrounding Sulfur's house there was a sea of yellow sunflowers. It looked like their smiling happy faces were welcoming him to Sulfur's property, #16, 6th Street. Professor Terry said, "No surprise, yellow is the family color."

As the bubble landed Sulfur was standing in front of his yellow house. Guy noticed Sulfur's 6e hair and knew that he was in Group 6A on the Periodic Table. The triangle on his body indicated that he was in Period 3. Guy said, "When I draw your Bohr model you will know what the Element Box has to do with your atom, and you'll know for sure that your body markings are correct."

Sulfur insisted on giving them a tour of his entire house. It was amazing. All the rooms and all the furniture were painted yellow. Guy thought Sulfur must really love his family color. At least it's a happy color.

Guy showed Sulfur the Steps Chart he had created to help him learn how to draw a Bohr model by himself. They decided to draw the Bohr model out on the yellow porch of his yellow house. Out there, they moved the Steps Chart close to the table where Guy would work. Sulfur had provided a large pad for Guy to draw on. With everything in place, Guy picked up his magic marker and began to draw Sulfur's Bohr model.

Guy Gets Busy Drawing Sulfur's Bohr Model

1. Label the Columns:

ELEMENT BOX CALCULATIONS BOHR MODEL

2. Copy and Label the Element Box from the Periodic Table.

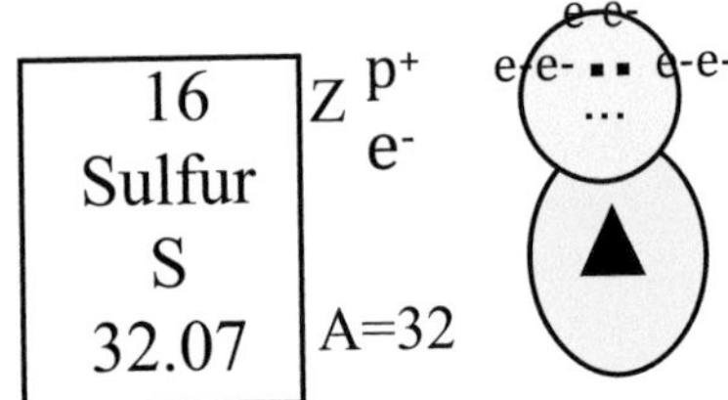

3. Do the Calculations

$n^0 = A - Z$

$n^0 = 32 - 16$

$n^0 = 16$

4. Draw Sulfur's Atom

1st draw a thick circle for the nucleus.

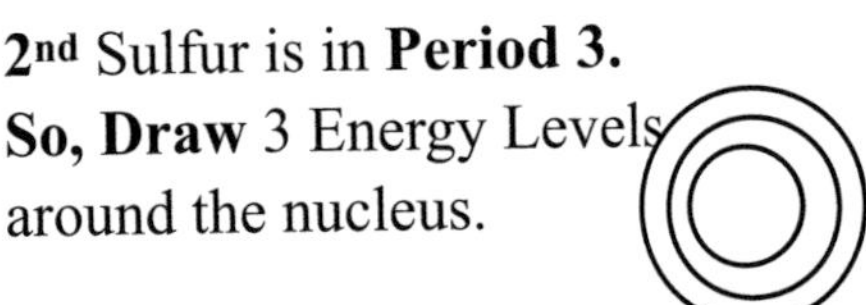

2nd Sulfur is in **Period 3.**
So, Draw 3 Energy Levels around the nucleus.

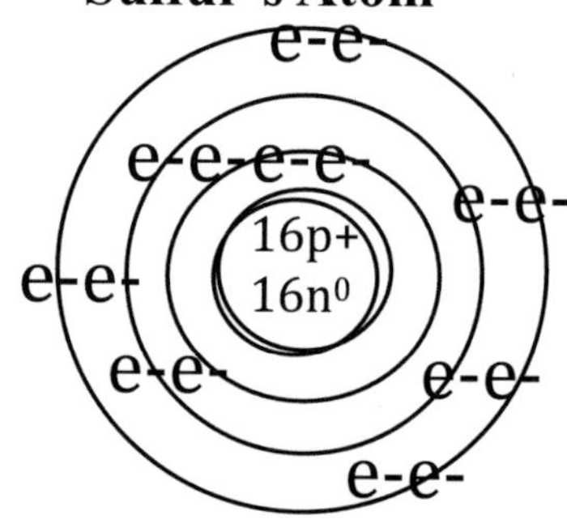

3rd Put the protons and neutrons in the nucleus.

p+ = Atomic #Sulfur's' is 16 16p+

n^0neutrons (See Calculations). $16n^0$

4th Put electrons in Energy Levels.

\# of Electrons equals the Atomic #.
Sulfur's Atomic Number is 16.
Put 16 electrons in the energy levels.

Guy said, "I've finished your Bohr model. Come look at it."

SULFUR'S BOHR MODEL

ELEMENT BOX CALCULATIONS BOHR MODEL

16
Sulfur
S
32.07

Z p^+ e^-

A=32

$n^0 = A - Z$

$n^0 = 32 - 16$

$n^0 = 16$

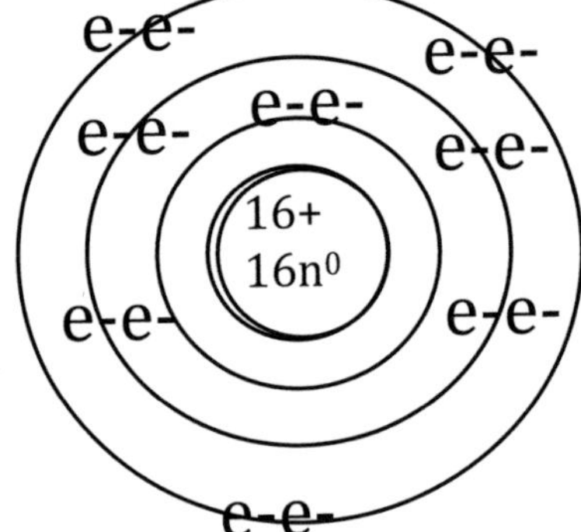

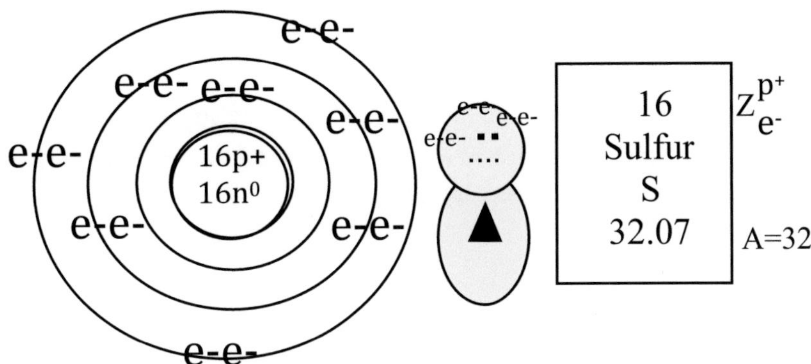

Sulfur looked at his Bohr model and said, "This is the first time I have seen my 3 energy levels. Period 3 told me I had them, but it's so good to see I really have 3 energy levels. I also heard that my Group was 6A because I had 6 electrons in my Outside Energy Level. Now I can actually see those 6 electrons. It is so exciting to see my atom. I was told that my Element Box described my atom, but here in my Bohr model I am seeing it. My Atomic Number is 16, and I do have 16 protons in my nucleus and 16 electrons in my energy levels. There they are. I can see them. It also describes my body markings. My triangle represents the 3 energy levels in my atom. My 6e hair symbolize the 6 electrons in my Outside Energy Level. I can actually see what both these body markings mean. Bohr models are so good."

SULFUR

Sulfur said, "Let's sit here on my porch where we can see the sun flowers. They make me happy. I'd like to tell you that in my element form I'm a yellow powder. I was made into medicine during the World War, and I have saved many lives since then. Some people don't like me when I join with hydrogen and make water smell like rotten eggs."

Guy said, "I guess people with infections would appreciate you, Sulfur."

Sulfur continued, "I form a very powerful acid called sulfuric acid which has uses in industry. I'm involved in the production of steel and rubber. I have been used as a bleach which is a way to kill fungus. I'm also used in the production of paper."

Guy said, "You sure have a lot of uses. I now know so much about the element Sulfur." Then they left, and Sulfur enjoyed looking at his Bohr model. He was happy.

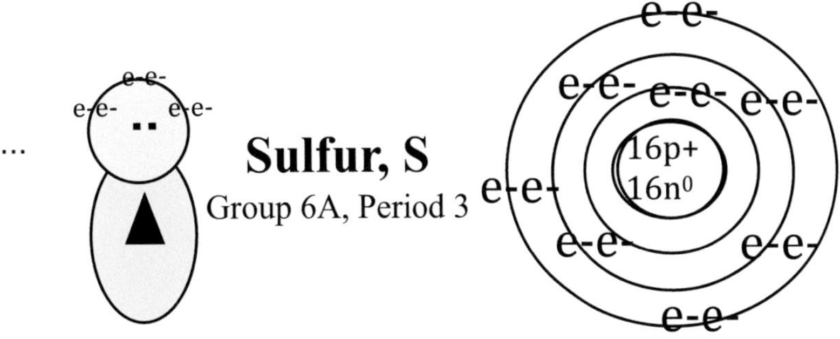

CHLORINE

Leaving Sulfur's yellow paradise, the magic bubble rose high above Periodic Table Land. 7th Street's Pine forest was in sight again. Looking beyond the Pine forest the scenery changed again. Guy said, "Look at all the swimming pools on 7th Street."

Professor Terry said, "It's not surprising. One of Chlorine's compounds is used to kill germs in swimming pools. Notice Chlorine's property where we are going. It looks like Chlorine has several pools. We'll land at the top of 7th Street behind the Pine forest, and stash the balloons and bubble behind that really old pine tree like we did before. When they landed, Guy remembered to color Sulfur's box yellow. Guy remarked, "Look how many element boxes we've colored so far. Only four more to go."

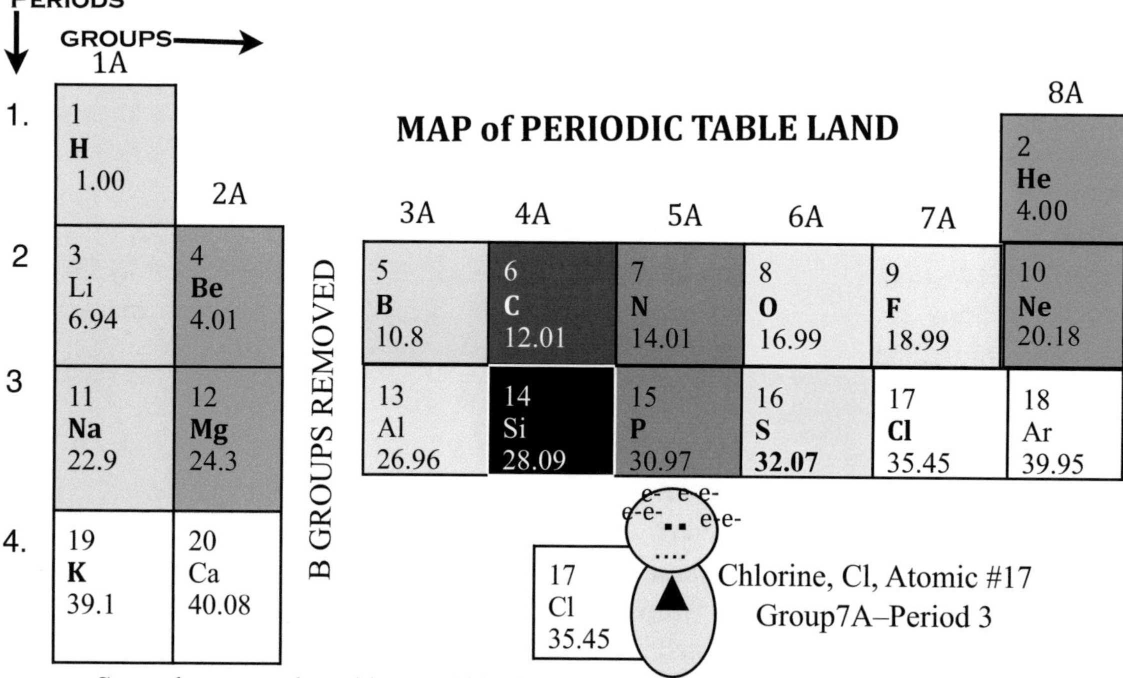

Soon they were knocking at Chlorine's door. Out popped Chlorine with his 7e hair and triangle on his body. The 7e hair told Guy that Chlorine was in Group 7A and the triangle showed he was in Period 3 on the Periodic Table.

Professor Terry explained that Guy was here to draw Chlorine's Bohr model. "It will be special to you. No other element will have one exactly like yours. It will show the meaning of everything about you that's in your Element box on the Periodic Table and the markings on your body–your 7 e hair and your triangle."

When Chlorine gave Guy and Professor Terry a tour of his house, Guy couldn't believe that Chlorine also had an indoor pool to enjoy if the weather was bad. Chlorine suggested his den as the best place to draw his Bohr model.

Guy took out the Steps Chart and explained that the information on that chart told step by step how to draw a Bohr model. "This will help you learn how to draw your own Bohr model some day." Guy said "Let's put the chart right near the desk here. That way you can look at it while I draw your Bohr model on this pad." Then Guy began.

Guy Gets Busy Drawing Chlorine's Bohr Model

1. Label the Columns:

ELEMENT BOX CALCULATIONS BOHR MODEL

2. Copy and Label the Element Box from the Periodic Table.

17 Chorine Cl 35.45

Z p^+ e^-

A=35

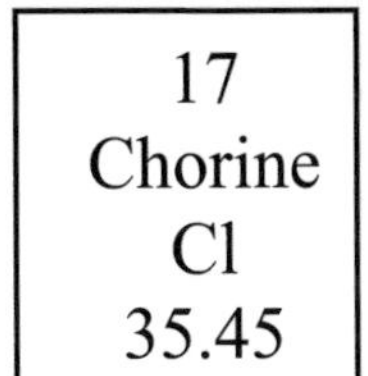

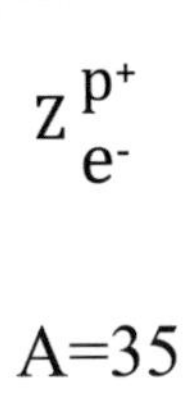

3. Do the Calculations

$n^0 = A - Z$

$n^0 = 35 - 17$

$n^0 = 18$

4. Draw Chlorine's Atom

1st draw a thick circle for the nucleus

2nd Draw energy levels. Chlorine is in Period 3
So, draw 3 energy levels.

3rd Put the protons and neutrons in the nucleus.
p+ = Atomic # Chlorine's is 17
n^0neutrons (See Calculations).

17 p+
18 n^0

Chlorine's Atom

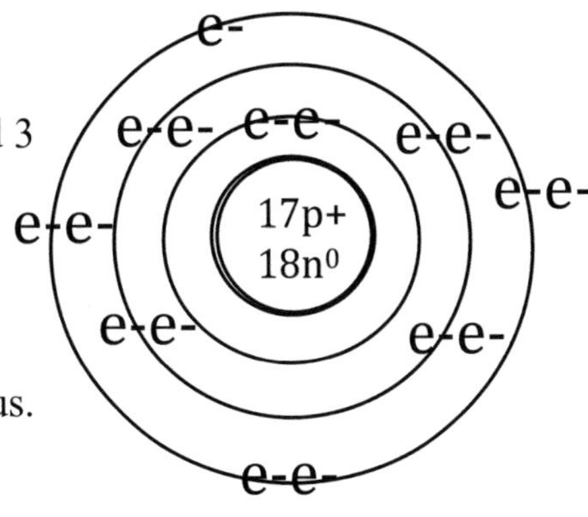

4th Put electrons in Energy Levels.

# of Electrons equals the Atomic Number.
Chlorine's Atomic Number is 17.
Put 17 electrons in the energy levels.

Guy said, "I've finished drawing your Bohr model.
Come look at it."

CHLORINE'S BOHR MODEL

ELEMENT BOX CALCULATIONS BOHR MODEL

17 Chorine Cl 35.45

Z p^+ e^-

A=35

$n^0 = A - Z$

$n^0 = 35 - 17$

$n^0 = 18$

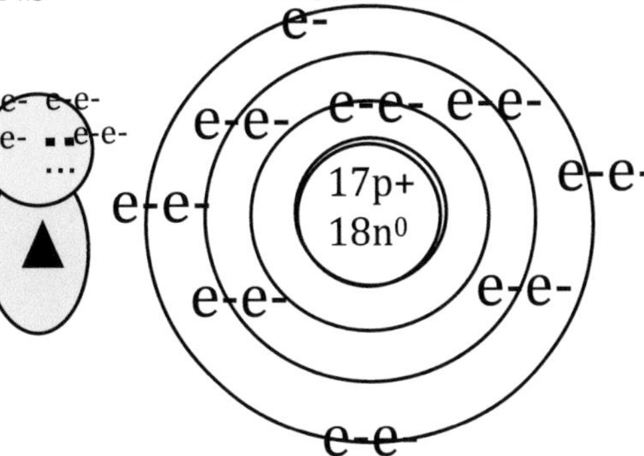

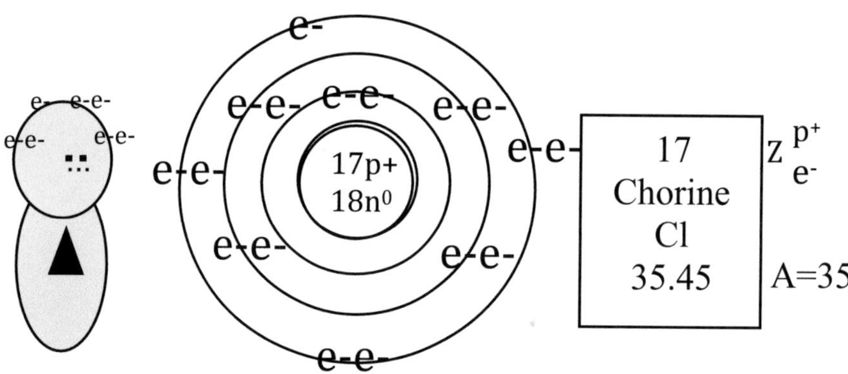

Chlorine looked, and the Bohr model surprised him with its correctness. “Oh my goodness, there are the 17 protons and 17 electrons in my atom, just like my Atomic Number said I should have. Wow! I see the electrons in my Outside Energy Level add up to 7, telling everyone I’m in Group 7A. So the 7e hair on my head are right when they tell that I have 7 electrons in my Outside Energy Level." Then he counted the energy levels in his atom and there were 3. "I guess being in Period 3 really means that my atom has 3 energy levels. Also, the triangle on my body which tells people at a glance that I have 3 energy levels is correct. My atom does have 3 energy levels. This is wonderful Guy. Thank you for teaching me about my Bohr model.”

CHLORINE

“Come over here and sit by my pool. I have so much to tell you about my uses.” Chlorine said, “In my normal state I’m a yellowish green poison gas which is banned from use in war. People add one of my products to their wash to bleach clothes. I’m used to kill germs in swimming pools and make them safe for people to swim. You can tell this is one of my favorite uses. As you see I have my own special swimming pools right on my property. Speaking of killing germs, I’m used to purify water to make it safe to drink. This is important because everyone needs to drink water, and if it’s not pure, people get sick. Next, one of my favorite compounds I form with Sodium is Sodium chloride. That’s ordinary table salt and it makes people’s food taste so good. They use me in the production of paper and cloth. With Hydrogen I make hydrochloric acid which has a long list of uses in industry and in chemistry labs.”

Chlorine loved his very own Bohr model. It filled his whole body with joy. He was so happy.

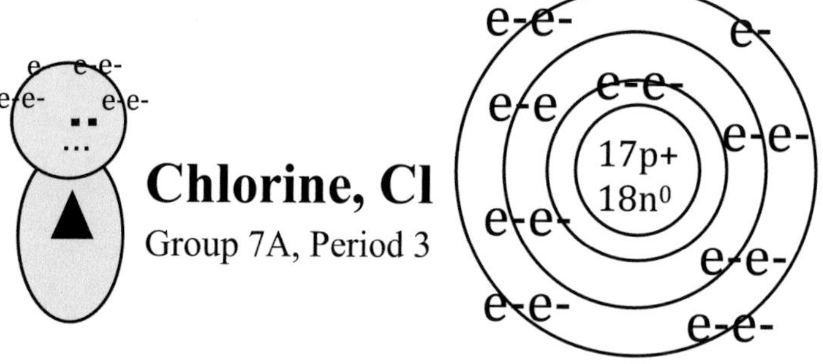

ARGON

Off they went to visit Argon on 8th Street. Before they climbed into the magic bubble Guy remembered to color Chlorine's box on the map of Periodic Table Land. The color was a yellow green, the chemical family color for elements in Group 7A.

MAP of PERIODIC TABLE LAND

PERIODS ↓ GROUPS →

Periods	1A	2A	B GROUPS REMOVED	3A	4A	5A	6A	7A	8A
1.	1 **H** 1.00								2 **He** 4.00
2	3 Li 6.94	4 **Be** 4.01		5 **B** 10.8	6 **C** 12.01	7 **N** 14.01	8 **O** 16.99	9 **F** 18.99	10 **Ne** 20.18
3	11 **Na** 22.9	12 **Mg** 24.3		13 Al 26.96	14 Si 28.09	15 **P** 30.97	16 **S** **32.07**	17 **Cl** 35.45	18 Ar 39.95
4.	19 **K** 39.1	20 Ca 40.08							

e- e- e- e- e- e- e- e-

18 **Ar** 18.95

Argon, Ar, Atomic #18
Group 8A–Period 3

On the way to 8th Street the sun performed its little miracle—a rainbow was making an arch over the houses. He recalled the little pneumonic that he learned to remember the colors of the rainbow in order. ROY G BIV—**R**ed, **O**range **Y**ellow **G**reen **B**lue **I**ndigo and **V**iolet. Seeing this rainbow reminded Guy of his love for his beautiful world—the reason he began to study chemistry. Professor Terry found a safe place to store their trusty means of transportation remembering to hang the ropes over a low branch to keep them from getting tangled. Then they walked past Neon's place to finally arrive at #18, the home of Argon. Since Argon is found in our atmosphere, his house was painted a sky blue color and decorated with fluffy white clouds. The fence which surrounded the house had posts topped with old light bulbs signifying one of the main uses for Argon—making light bulbs last longer.

Argon came out and greeted Guy and Professor Terry. She announced that Guy had come to draw Argon's Bohr model. Guy pointed out that the meaning of Argon's 8e hair and the triangle on his body, would be verified along with his information on the Periodic Table when he saw his Bohr model.

Guy showed Argon the Steps Chart that he had created to teach the elements how they could draw their own Bohr model. Guy said, "We'll put this chart where you can read what it says while I draw your Bohr model." Argon showed Guy around his house ,and they found the perfect place to set up and draw Argon's Bohr model. Guy began.

Guy Gets Busy Drawing Argon's Bohr Model

1. Label the Columns:

ELEMENT BOX CALCULATIONS BOHR MODEL

2. Copy and Label the Element Box from the Periodic Table.

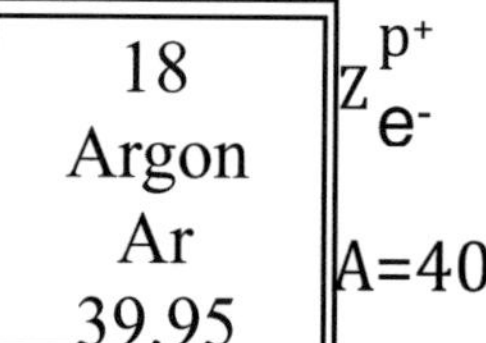

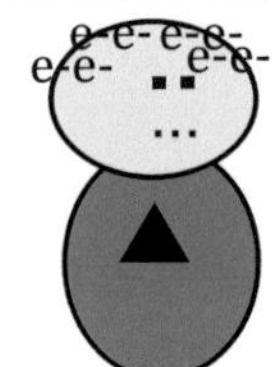

3. Do the Calculations

$n^0 = A - Z$

$n^0 = 40 - 18$

$n^0 = 22$

4. Draw Argon's Atom

1st draw a thick circle for the nucleus

Argon's Atom

2nd Draw energy levels. Argon is in Period 3
So, draw 3 energy levels

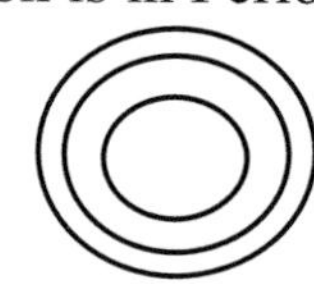

3rd Put the protons and neutrons in the nucleus.

p+= Atomic # Argon's is 18

n^0 neutrons (See Calculations).

18 p+

22 n^0

e-e- e-e- e-e- e-e- e-e- e-e- e-e- e-e- e-e-

18p+

$22n^0$

4th Put electrons in Energy Levels. How many? It equals the Atomic Number. Argon's Atomic Number is 18. So put 18 electrons in the energy levels.

Guy said, "I've finished drawing your Bohr model. Come look at it."

ARGON'S BOHR MODEL

ELEMENT BOX	CALCULATION.	BOHR MODEL

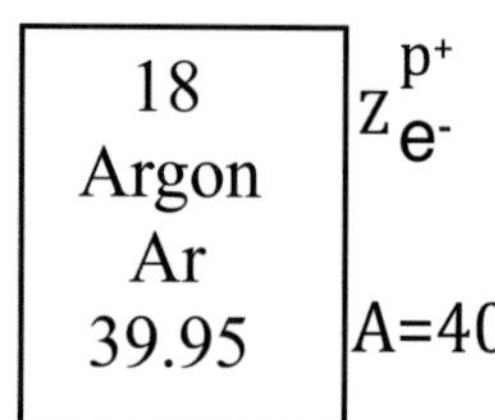

$n^0 = A - Z$

$n^0 = 40 - 18$

$n^0 = 22$

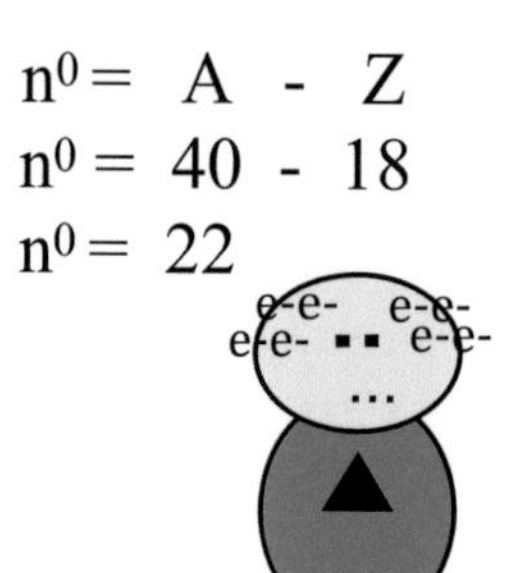

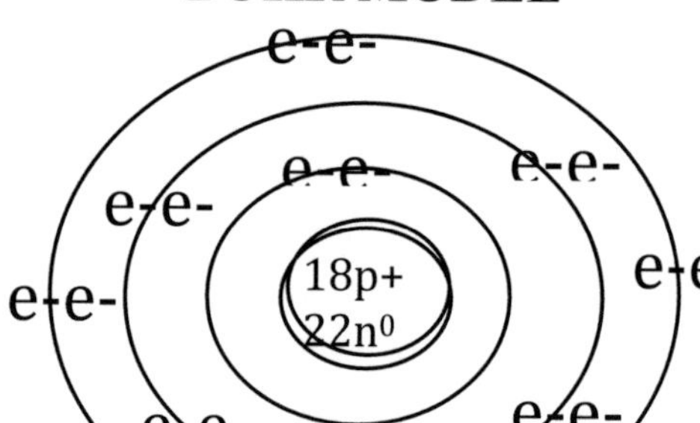

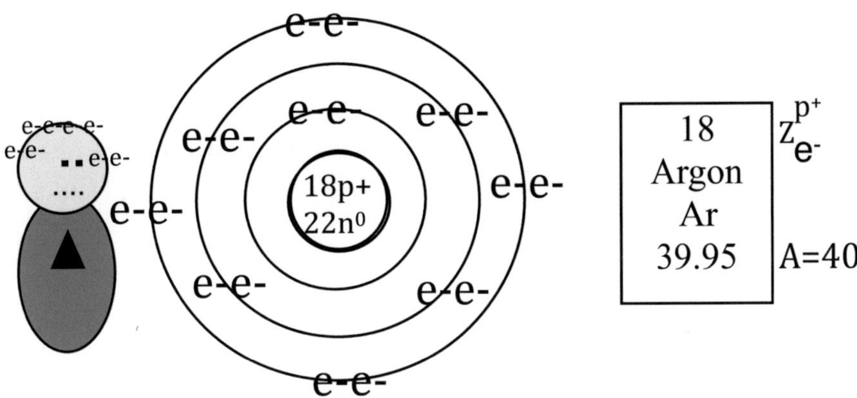

Argon took a close look at his Bohr model and was excited. He knew he was in Period 3. His body triangle said so. "How great to see the 3 energy levels right there in my atom! I see 18 electrons in my energy levels and 18 protons in my nucleus. My Atomic Number, 18, in my Element Box had told me that. Now I know it's true." Then as Professor Terry had requested, he checked out his body markings with the Bohr model. "My 8e- hair says I have 8 electrons in my atom's Outside Energy Level. So does being in Group 8A. My Bohr model really has 8 electron in its Outside Energy Level." He counted them. "Everything checks out. It's just great. Thank you for showing me how to draw my Bohr Model."

ARGON

Argon said, "I'm element Atomic Number 18 and one of the Noble Gases. I make up 1% of the earth's atmosphere. Guy approached him and asked what he was used for. Argon said, "I am added to light bulbs to keep the filament from being corroded by Oxygen. That way I prolong the life of the bulb. If you run an electric current through me in a tube I turn a beautiful sky blue color. I'm used as an arc gas shield by welders to protect the weld area. I also make a beautiful blue-green gas laser.".

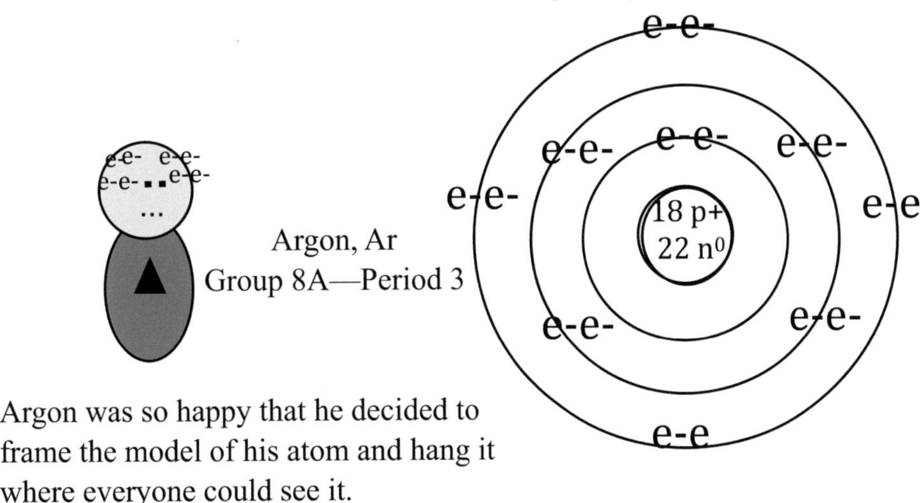

Argon was so happy that he decided to frame the model of his atom and hang it where everyone could see it.

POTASSIUM

After remarking about how many uses Argon had, Guy thought about coloring Argon's box a royal purple like Helium and Neon. Then he said, "We are almost finished drawing Bohr models——only two more elements to visit, Potassium and Calcium "

MAP of PERIODIC TABLE LAND

PERIODS ↓ GROUPS →

B GROUPS REMOVED

Periods	1A	2A	3A	4A	5A	6A	7A	8A
1.	1 **H** 1.00							2 **He** 4.00
2	3 Li 6.94	4 **Be** 4.01	5 **B** 10.8	6 **C** 12.01	7 **N** 14.01	8 **O** 16.99	9 **F** 18.99	10 **Ne** 20.18
3	11 **Na** 22.9	12 **Mg** 24.3	13 Al 26.96	14 Si 28.09	15 **P** 30.97	16 **S** **32.07**	17 **Cl** 35.45	18 Ar 39.95
4.	19 **K** 39.1	20 Ca 40.08						

19 **K** 39.09

e-

Potassium, K, Atomic #19
Group 1A–Period 4

Professor Terry said, "Let's climb into the bubble now and color the map while we're traveling". The green balloons did their job when Guy pointed the magic wand in the direction of 1st Street. Guy felt the refreshing cool breeze blowing through his hair. Coloring Argon's box on the map, made the trip to 1st Street seemed to pass by so much faster than usual. Soon they were landing on the cleared space behind 1st Street.

In no time at all they found their way down to #19, 1st Street and found themselves knocking at Potassium's door. When Potassium answered the door, Guy noticed the 1 electron he had for hair and the diamond shape on his body. Guy said,"When you get to see your Bohr model you will be so happy. You will see that your body markings do mean what you have always believed. Your 1e hair says you're in Group 1A. The diamond with its 4 sides says you are in Period 4 with 4 energy levels."

Potassium said, "Lithium and Hydrogen both told me you had drawn their Bohr models. Other elements told me I would better understand the information in my Element Box on the Periodic Table as well as my body markings."

Guy showed Sodium the Steps Chart he had created saying, "Here's a chart that will make it easier for you to learn how to draw your own Bohr model. You can read it and follow along as you watch me draw your Bohr model." Potassium gave Professor Terry and Guy a tour of his whole house. Professor Terry said, "Here's a great place to draw your Bohr model." It was Potassium's solarium. The air was warm and moist with perfect lighting. Then Guy began to draw Potassium's Bohr model.

Guy Gets Busy Drawing Potassium's Bohr Model

1. Label the columns:

ELEMENT BOX CALCULATIONS BOHR MODEL

2. Copy and Label the Element Box from the Periodic Table.

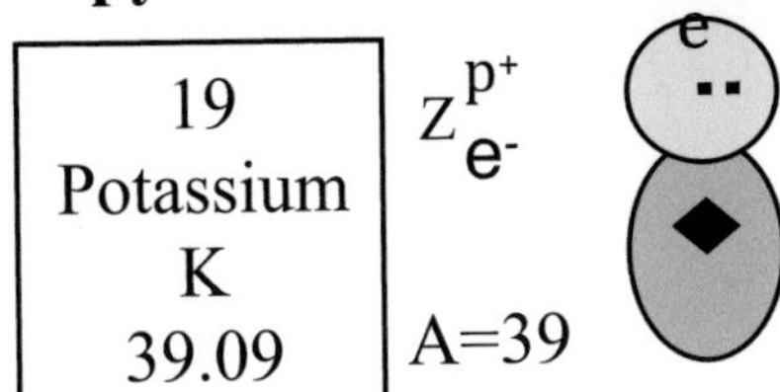

3. Do the Calculations

$n^0 =$ A - Z

$n^0 =$ 39 - 19

$n^0 =$ 20

4. DrawPotassium's Atom

Potassium's Bohr Model

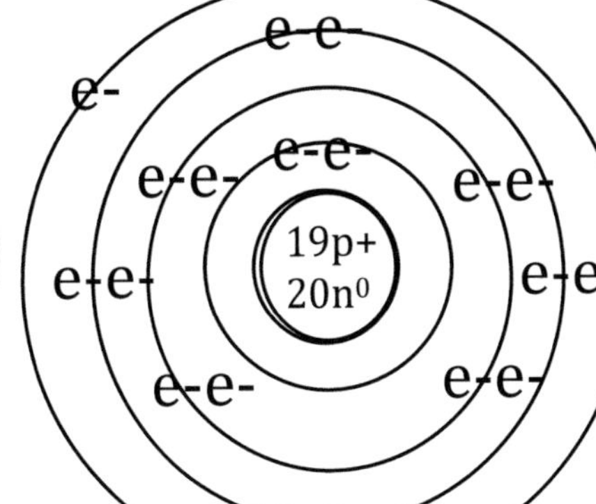

1st draw a thick circle for the nucleus

2nd Draw energy levels. Potassium is in Period 4 So, draw 4 energy levels.

3rd Put the protons and neutrons in the nucleus.

p+= Atomic # Potassium's is 19 19p+

n^0 neutrons (See Calculations $20n^0$

4th Put electrons in Energy Levels. How many? It equals the Atomic Number. Potassium's Atomic Number is 19. So put 19 electrons in the energy levels.

Guy said, "I've finished drawing your Bohr model. Come look at it."

POTASSIUM'S BOHR MODEL

ELEMENT BOX CALCULATIONS BOHR MODEL

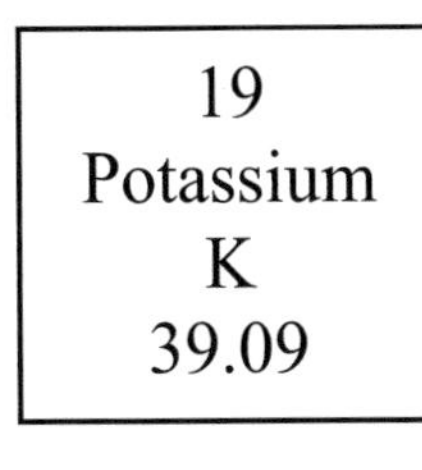

Z p+ e-

A=39

$n^0 =$ A - Z

$n^0 =$ 39 - 19

$n^0 =$ 20

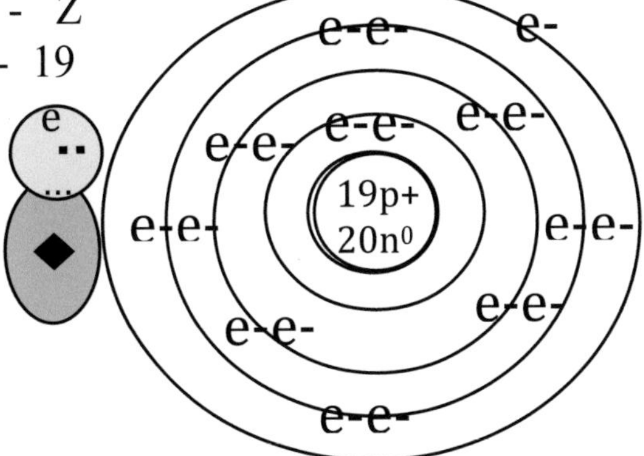

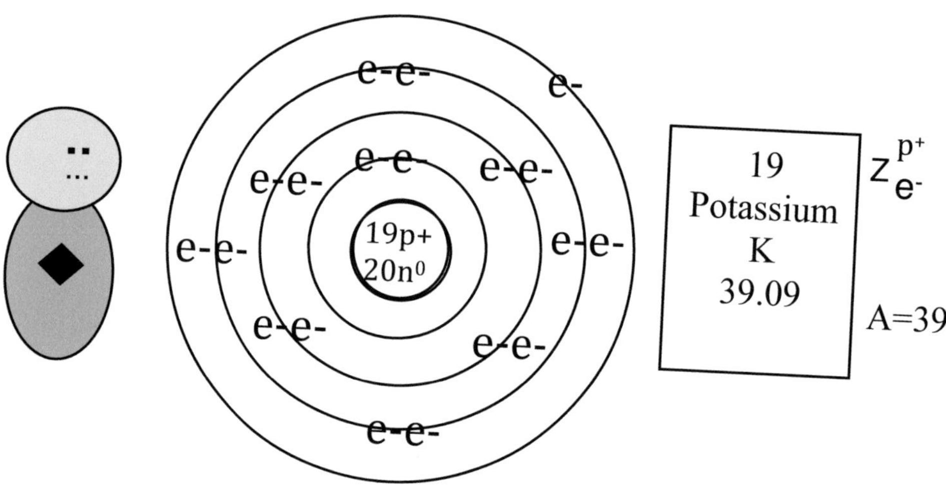

Potassium looked carefully at his Bohr model. "My Atomic Number 19 says I have 19 protons in my nucleus and 19 electrons in my energy levels. There they are. In the nucleus I see the 20 neutrons that I calculated using information from my Element Box. The Periodic Table told me that I was in Period 4. Now I can see that my atom has 4 energy levels. That's what the diamond on my body also showed everyone and it's true. I've been told I was in Group 1A on the Periodic Table, my 1e hair says that too. This means I have 1 electron in my Outside Energy Level and I do. My Bohr model is fantastic. It shows everything that the Periodic Table and my body markings have always said would be in my atom. Wow!"

POTASSIUM

Potassium said, "Let's sit here and visit. I'm a soft silvery white metal, and it's my compound, Potassium phosphate that replenishes the soil on farms when the crops use up its nutrients. Best of all, I work with Sodium to keep your heart beating properly. Some people with high blood pressure can't use regular salt. Potassium chloride, KCl is a good salt substitute. Potassium carbonate is used to make specialty glass like TV tubes Potassium hydroxide is used to make soaps, detergents and drain cleaners."

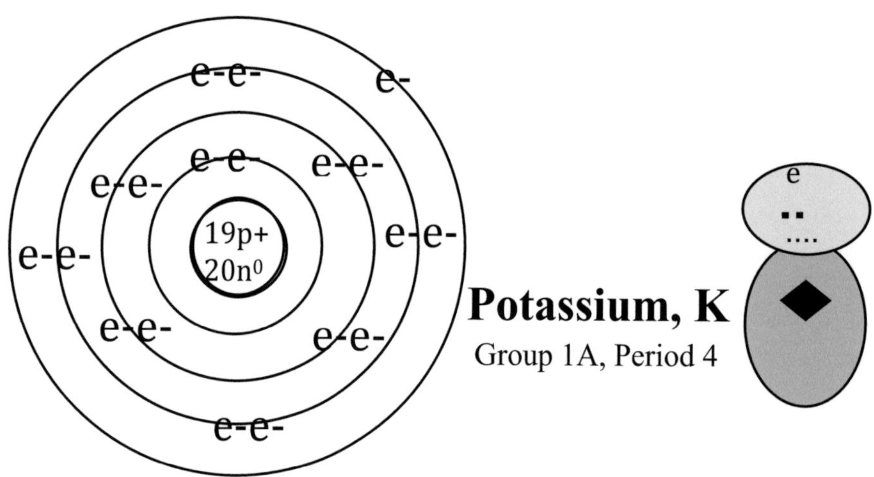

CALCIUM

After they left Potassium, Professor Terry reminded Guy that they had to color in Potassium's box on the map. Since Potassium was in Group 1A with Sodium, Lithium and Hydrogen, it was easy to see that the box needed to be colored a light green, obviously the family color. That job taken care of they hopped into the magic bubble and they were off to see Calcium in Group 2A. Calcium's home was at #20, 2nd Street.

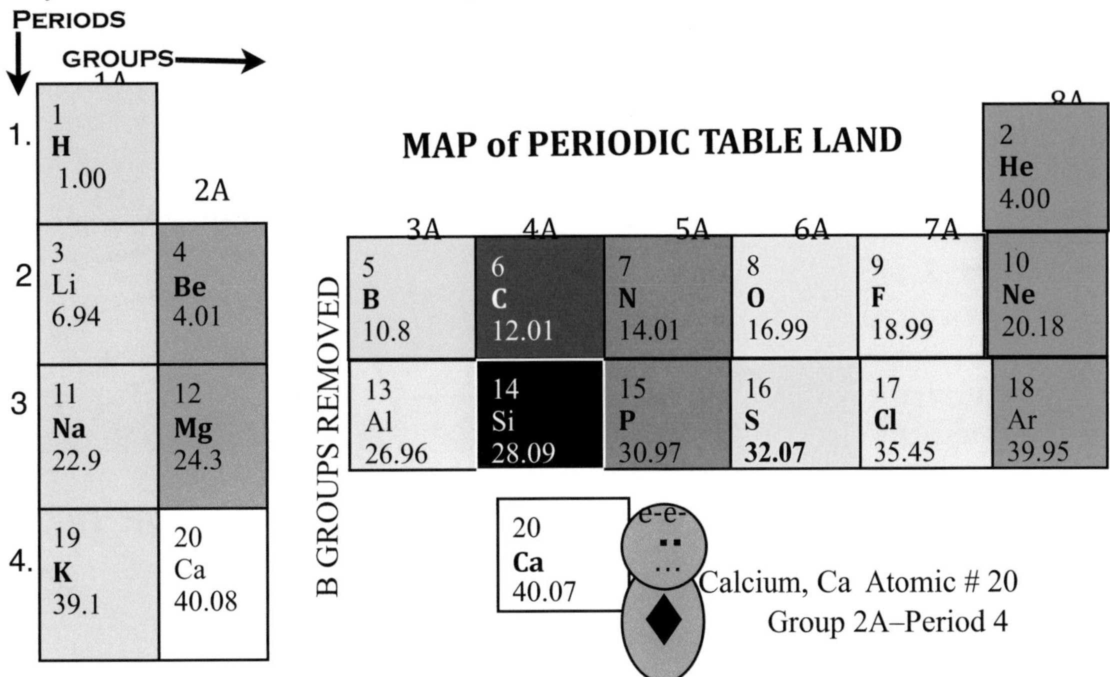

It was a quick flight. The light green balloons took them so high into the sky that the houses looked miniature. Guy loved the wonderful view they had of all of Periodic Table Land. Guy looked in every direction and remembered drawing the Bohr models for all of those elements, Atomic Numbers 1 to 19. Calcium #20 was the last to be drawn.

After landing, they moved quickly on to Calcium's home. They knocked at the door, and out came Calcium. Guy quickly noticed his 2e hair and the four sided diamond on his body. Guy remembering the baseball field's diamond. said, "Once my father took me to a National League baseball game. The shape of the field was called a diamond. It had 4 sides like the figure on your body.

Calcium invited them into his home. Professor Terry and Guy were friends with Calcium. Sitting down Guy said, "Well, I've come to draw your Bohr model. When I do, you will be able to see what your atom looks like. You will love it. I know that because you are the last element that I will draw and every single element has. It will verify the meaning of your body markings and also the meaning of the information in your Element Box."

"Here's a chart that I made up for you to show the steps to follow in creating your Bohr model. If you watch the Steps Chart while I am drawing your atom, it will help you learn how to do it yourself some day." Calcium had a sunny screened in porch where he had a desk set up for Guy. Calcium hung the Steps Chart near the desk, pulled a chair close, and he was ready to learn how to draw a Bohr model. Guy began.

Guy Gets Busy Drawing Calcium's Bohr Model

1. Label the columns:

ELEMENT BOX CALCULATIONS BOHR MODEL

2. Copy and Label the Element Box from the Periodic Table.

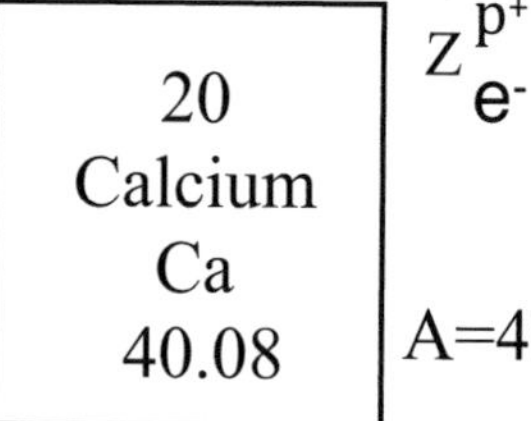

3. Do the Calculations

$n^0 = A - Z$

$n^0 = 40 - 20$

$n^0 = 20$

Calcium's Atom

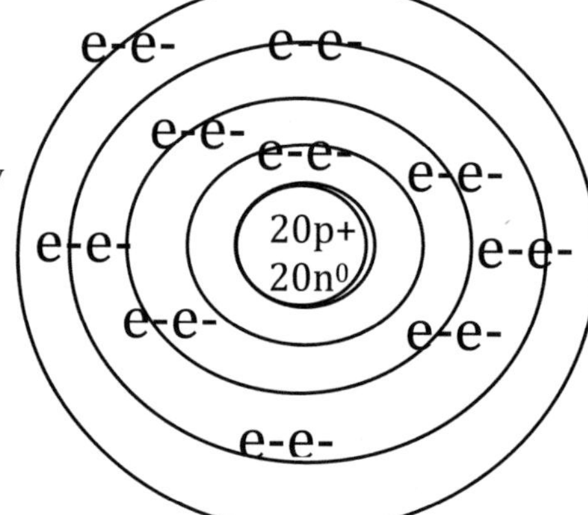

4. Draw Calcium's Atom1

1st draw a thick circle for the nucleus. 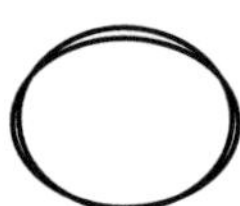

2nd Calcium is in **Period 4. So, Draw** 4 Energy Levels around the nucleus.

3rd Put the protons and neutrons in the nucleus.

p+ = Atomic # Calcium's is 20 20p+

n^0neutrons (See Calculations $20\ n^0$

4th Put electrons in Energy Levels.

# of Electrons equals the Atomic Number.

Calcium's Atomic Number is 20.

Put 20 electrons in the energy levels.

Guy said,"I've finished drawing your Bohr model. Come look at it.

CALCIUM'S BOHR MODEL

ELEMENT BOX CALCULATIONS BOHR MODEL

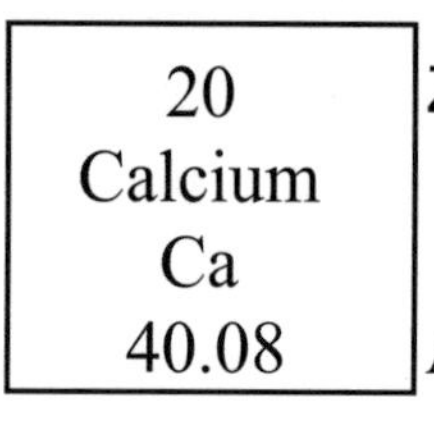

$n^0 = A - Z$

$n^0 = 40 - 20$

$n^0 = 20$

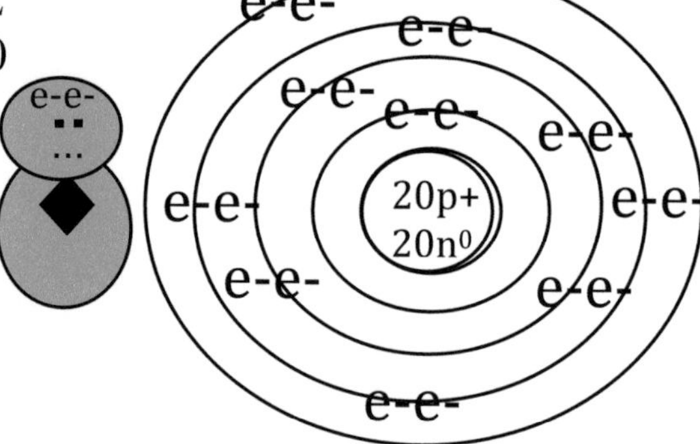

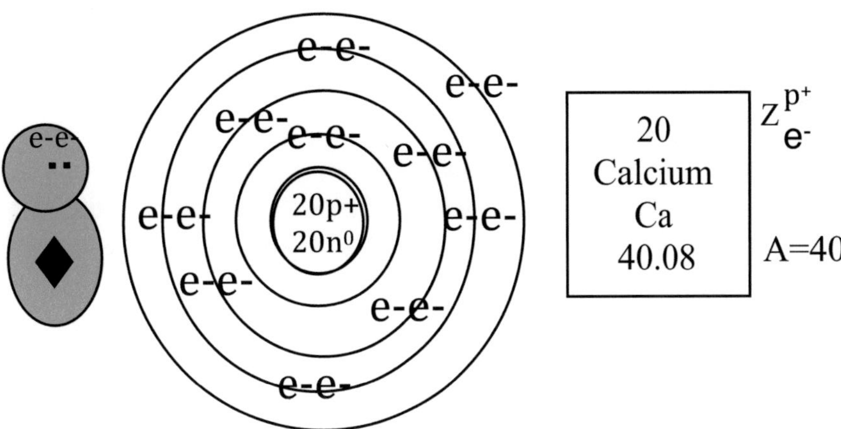

Calcium saw how well his Bohr model displayed who he was. He looked and saw 20 p+(protons) in his nucleus and 20 e- (electrons) in his energy levels. It was just as his Atomic Number said there should be, 20. He took an extra look at his 20 protons and said proudly, "These 20 protons make me, me! No other element in the whole universe has 20 Protons in its nucleus." There were 4 energy levels in his atom, just as Period 4 said there would be. "Since I'm in Group 2A, there should be 2 electrons in my Outside Energy Level." Calcium checked, and there they were. Calcium found everything else as it should be. One look at his Bohr model told him a lot. "What my Bohr model says makes my body markings correct too. It's amazing!"

CALCIUM

Calcium said, "Sit down now and learn something about me. I've been around since the time of the ancient Greeks and Romans. I've always been part of the earth in limestone rock and under the sea in beautiful coral reefs. You'll find me in shells of clams and other sea creatures, as well as in teeth and bones. I play an important role in the body. I help your muscles to squeeze and release, your blood to clot, your heart to beat properly, and your nerves to transmit messages. To keep these functions going smoothly, it's important to have Calcium in your diet. So eat food with Calcium in it.

Guy was impressed hearing that Calcium was an important part of the human body in so many different ways. He would try to find out what foods containe Calcium.

Calcium said, "I have more to tell you. "I'm used in construction as part of cement and wall boards. I'm also appreciated in areas that have snow and ice. My compound calcium chloride melts ice to keep people from slipping and falling. At airports I'm used to deice planes."

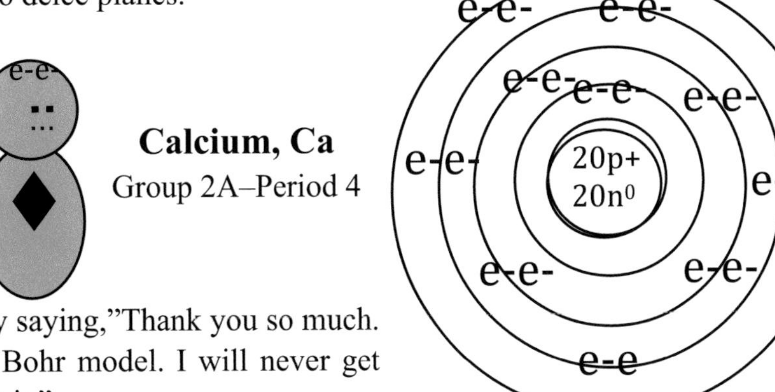

Calcium, Ca
Group 2A–Period 4

Calcium ended by saying,"Thank you so much. I really love my Bohr model. I will never get tired of looking at it."

Bohr Models Completed

Guy was finished drawing Bohr models. It was about time to leave Periodic Table Land. "Let's complete this section on Bohr models by coloring the box for the very last element on the Map of Periodic Table Land" suggested Professor Terry.

Guy said, "The last element I drew a Bohr model for was Calcium". So he got busy and filled in Calcium's box brown like the other elements in Group 2A.

PERIODS ↓ **GROUPS** →

MAP of PERIODIC TABLE LAND

Periods	1A	2A	B GROUPS REMOVED	3A	4A	5A	6A	7A	8A
1.	1 **H** 1.00								2 **He** 4.00
2	3 Li 6.94	4 **Be** 4.01		5 **B** 10.8	6 **C** 12.01	7 **N** 14.01	8 **O** 16.99	9 **F** 18.99	10 **Ne** 20.18
3	11 **Na** 22.9	12 **Mg** 24.3		13 Al 26.96	14 Si 28.09	15 **P** 30.97	16 **S** **32.07**	17 **Cl** 35.45	18 Ar 39.95
4.	19 **K** 39.1	20 Ca 40.08							

Professor Terry said, "Now that you have finished Bohr models, I have a quiz for you to see if you have learned the most important principles the Bohr Model reveals. I will list what the Bohr model tells us about the element. You tell me the information about this element below and which parts of the Bohr model give us this information.

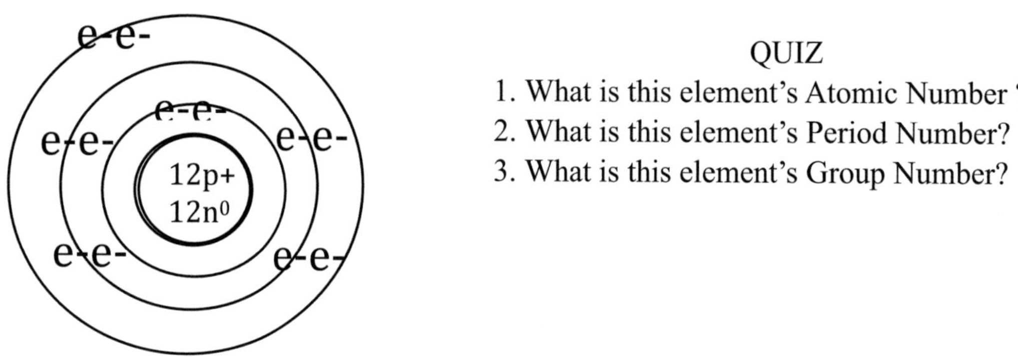

QUIZ

1. What is this element's Atomic Number ?
2. What is this element's Period Number?
3. What is this element's Group Number?

Ans: **1**. The **Atomic # is 12** Count the # of electrons in the energy levels and see the # of protons in the nucleus. **2. The Period # is 3** Count the number of energy levels. **3. Group 2A** Count the number of electrons in the Outside Energy Level

Professor Terry said, “If you don’t understand the answers to this Quiz, you need to select any one element from 3 to 20 and read the page after drawing the Bohr model for that element. You will see the Bohr model explained there.”

“Moving on, Guy. You noticed that we colored each element in the same A Group the same color. Soon you will know why. Each Group is the home of a different Chemical Family. In Periodic Table Land we use color to distinguish between the different Chemical Families. You are going to learn about the Chemical Families next. You need to meet the elements in these families because you will meet them again as you learn more chemistry. Knowing the elements will be important as you learn about compounds.

Yes Guy, I’m getting you ready to learn how compounds are formed. I guess you remember how much you had to learn before you were able to draw Bohr models. So you know you have to be patient and learn some basics before you can understand how compounds are formed. That’s next.”

Guy suggested, “Let’s find a place to sit and talk before leaving.

Congratulations Guy! You look really happy, and you should be. I remember what started you on this adventure to learn Chemistry. It was your fascination with the beauty and wonders of our universe. Those twenty Bohr models of the atoms you have just drawn are some of the basic elements that make up the universe you love. I’m happy for you.”

Guy was now ready to share with Professor Terry what was on his mind. So he began. “As I was drawing the Bohr models and looking at these atoms, a thought occurred to me. I couldn’t help but notice that atoms I drew reminded me of our planetary system. The electrons whirl around the nucleus of the atom the way the planets revolve around the sun. I’m amazed at how much they are alike.”

“Well Guy, I can understand what you were thinking. However, I have to tell you, they are not exactly alike. There is a basic similarity between the atom and the planetary system. Let me say this. If our planetary system and the atom were works of art, I would be sure that the same artist had created both.”

Guy thought, similar but not the same. Well, they sure are like each other. Then Guy shared, “I’m a little sad that drawing Bohr models has come to an end. However, now that I’m finished, I can look at all the beautiful parts of my world and imagine them made up of tiny little atoms that are just like the Bohr models I just drew. I have learned so much. That makes me happy.”

****

Guy went back to his family’s cabin trying to imagine what Professor Terry had in store for him. He loved creating Bohr models. What could be better than that. Guy’s mind was spinning thinking about Chemical Families. Guy told his parents about his Bohr

Model adventure. They were delighted that Guy was not only learning Chemistry but he was having fun as he learned.

That night when it was finally dark enough, Guy walked up that familiar trail to his favorite spot on the mountain. He found the rock that was inclined to exactly the right angle. He sat and rested his head on the rock. The dark sky was filled with a thousand pinpoints of light. Then he spotted a shooting star.

In an instant Guy's old friend Wish Star came tumbling out of the sky. He landed in the pine tree nearby causing the tree's top branches to sway back and forth. Then Wish Star jumped down and sat right beside him. Wish Star looked at Guy and smiled.

"Oh Wish Star, it's so good to see you again. I have so much to tell you. I just finished learning what the atoms of the elements look like. I drew the Bohr models of the first 20 elements It was so fascinating seeing the atoms of the elements. Thank you so much for introducing me to Professor Terry. Her magic Periodic Table let me slide down a mirrored tunnel into Periodic Table Land. I flew around that land in a magic bubble. I'm having such a great summer. The best is how much chemistry I'm learning."

Wish Star said, "I just popped in to let you know I haven't forgotten you. I'm watching you from above. Just know I'm always here for you Guy if you ever need me for any reason." They sat for a while chatting about things. Wish Star told Guy how proud he was about how much he was learning. Finally he told Guy it was time to leave. Wish Star flew off into the night sky. Just as he had done before, he turned one last time, winked at Guy, waved good bye, and off he flew into the night sky.

Guy walked back to the cabin so happy. All the way back he was thinking of his visit with Wish Star. Then his mind moved ahead. He wondered how many more adventures he would have and what they would be like. Finally reaching the cabin, he slipped into his comfortable bed and fell asleep. He began dreaming happy dreams of atoms and all the little elements that had become his friends this summer.

BOOK 2
PART 2
THE CHEMICAL FAMILIES
Guy Meets Members of the Chemical Families During the Flag Festival

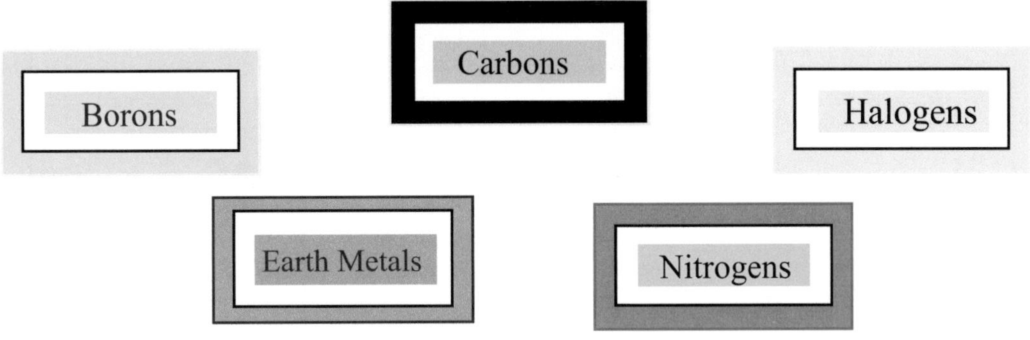

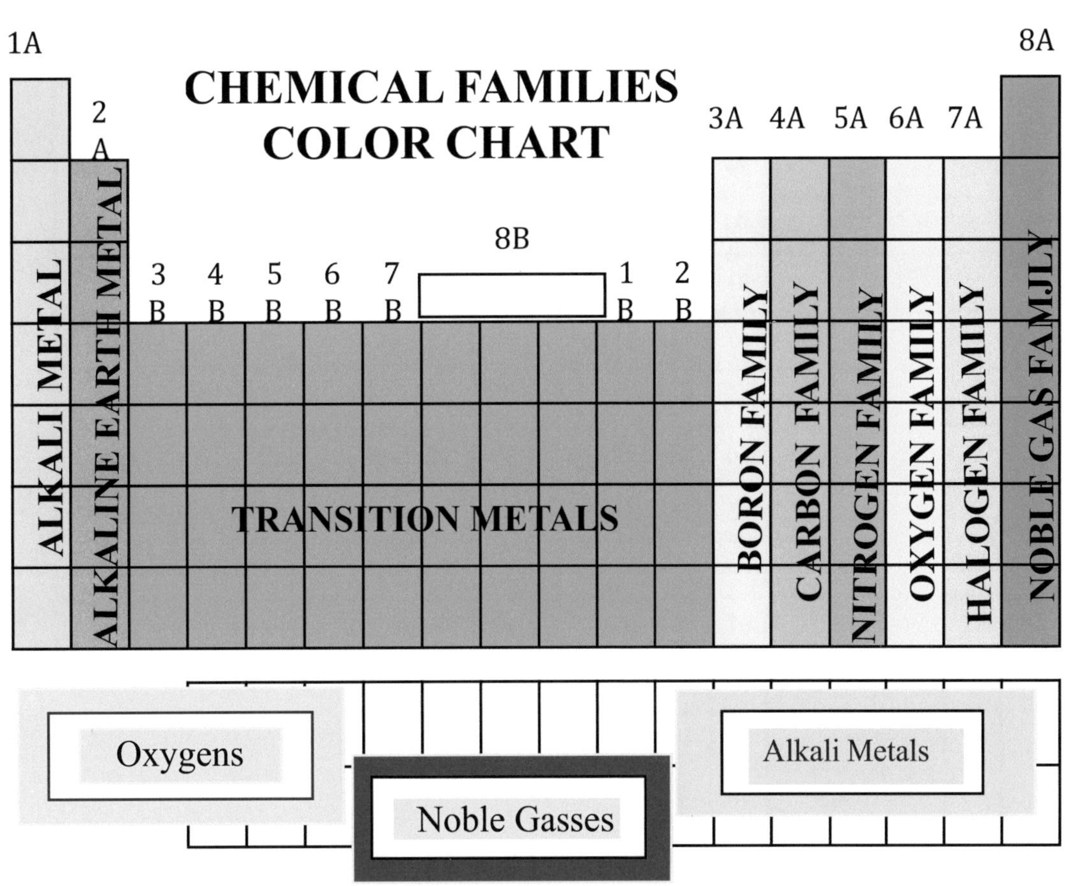

Charts Needed for Book 2, Part 2

See Page 16 for Elements 1 to 20

ELEMENTS /ATOMIC NUMBERS 21 TO 118 CHART

21	Scandium	47	Silver	73	Tantalum	99	Einsteinium
22	Titanium	48	Cadmium	74	Tungsten	100	Fermium
23	Vanadium	49	Indium	75	Rhenium	101	Mendelevium
24	Chromium	50	Tin	76	Osmium	102	Nobelium
25	Manganese	51	Antimony	77	Iridium	103	Lawrencium
26	Iron	52	Tellurium	78	Platinum	104	Rutherfordium
27	Cobalt	53	Iodine	79	Gold	105	Dubnium
28	Nickel	54	Xenon	80	Mercury	106	Seaborgium
29	Copper	55	Cesium	81	Thallium	107	Bohrium
30	Zinc	56	Radium	82	Lead	108	Hassium
31	Gallium	57	Lanthanum	83	Bismuth	109	Meitnerium
32	Germanium	58	Cerium	84	Polonium	110	Darmstadtium
33	Arsenic	59	Praseodymium	85	Astatine	111	Roentgenium
34	Selenium	60	Neodymium	86	Radon	112	Copernicium
35	Chlorine	61	Promethium	87	Francium	113	Nihonium
36	Argon	62	Samarium	88	Radium	114	Flerovium
37	Rubidium	63	Europium	89	Actinium	115	Moscovium
38	Strontium	64	Gadolinium	90	Thorium	116	Livermorium
39	Yttrium	65	Terbium	91	Protactinium	117	Tennessine
40	Zirconium	66	Dysprosium	92	Uranium	118	Oganesson
41	Niobium	67	Holmium	93	Neptunium		
42	Molybdenum	68	Erbium	94	Plutonium		
43	Technetium	69	Thulium	95	Americium		
44	Ruthenium	70	Ytterbium	96	Curium		
45	Rhodium	71	Lutetium	97	Berkelium		
46	Palladium	72	Hafnium	98	Californium		

BOOK 2
PART 2
The Chemical Families

TABLE OF CONTENTS

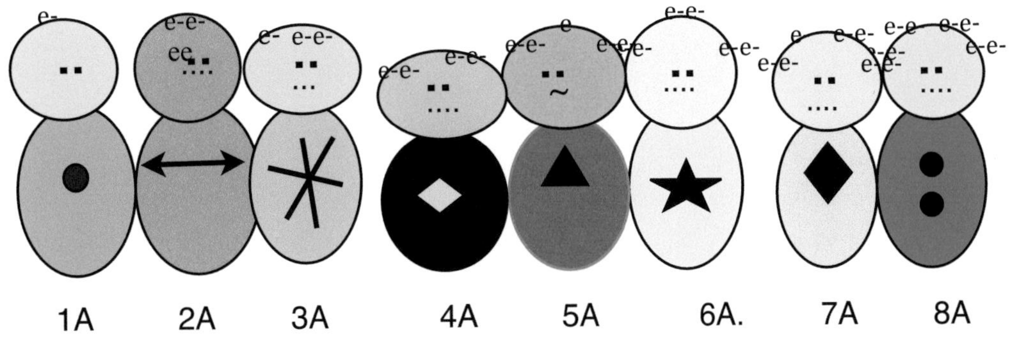

PART 2
Guy Gets Ready for his Chemical Family Adventure
Chapter 1

It was early and Guy was still sleepy. The bed felt so good as he rolled over for the third time. A little birdie hopped onto his window sill and chirped, "Wake up you little sleepy head. It's time to get out of bed." A cool breeze blowing across Guy's face and the aroma of pine and honeysuckle reminded him of his magnificent world that he was so eager to learn more about. It was the reason he wanted to learn chemistry, the science that explained the hidden world beneath the one he could see and love. "I've had a good start already understanding the Periodic Table and learning to draw the Bohr models of the elements. I'm thrilled to know my beautiful world is made of these same little atoms."

Finally, Guy got up, pulled on his clothes, enjoyed a great breakfast with his parents and made his way down the mountain trail to the lab. On the path down the mountain he noticed little chipmunks running in and out of the stone walls. There were wild flowers too on the side of the road. All these were part of the world he loved. Finally he arrived at the lab.

Professor Terry greeted Guy with news she knew would make him happy. "We're going back to Periodic Table Land to meet the Chemical Families. "Here's a chart I made for you showing the names of the Chemical Families."

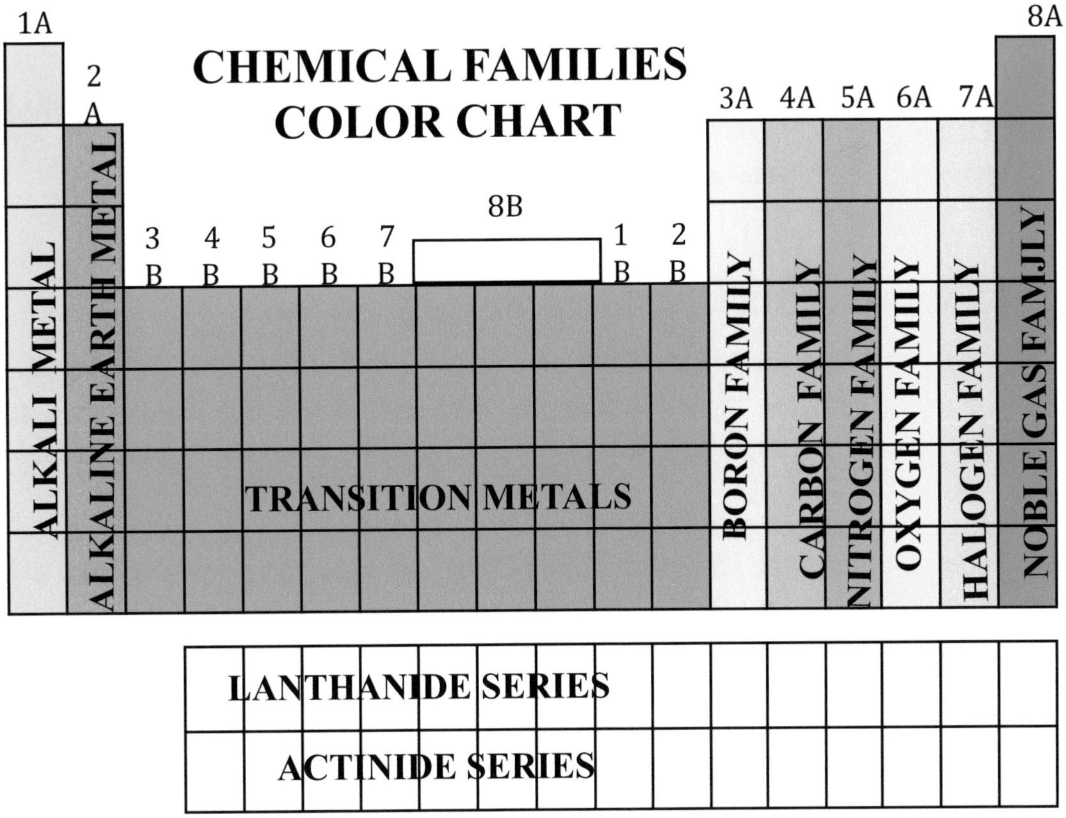

CHEMICAL FAMILIES COLOR CHART

1A
2A
3B 4B 5B 6B 7B 8B 1B 2B
3A 4A 5A 6A 7A 8A
ALKALI METAL
ALKALINE EARTH METAL
TRANSITION METALS
BORON FAMILY
CARBON FAMILY
NITROGEN FAMILY
OXYGEN FAMILY
HALOGEN FAMILY
NOBLE GAS FAMILY
LANTHANIDE SERIES
ACTINIDE SERIES

Professor Terry wanted Guy to understand what his chart meant. "Over in Periodic Table Land the little elements come in different colors—the color of their Chemical Family. My chart shows the colors of each of the Chemical Families Most elements are all one color. Here's a picture of Neon and he's purple, his family color. However we made his face a lighter color so you could see his features. The Boron family uses a light gray for their faces for the same reason."

"Over in Periodic Table Land purple is the color of all the elements in Group 8A on the Periodic Table. See Group 8A on my chart. It's purple. The Noble Gas Family chose purple for their color because purple is known to be a royal color. It's rather a fitting color for a noble family. Neon is a member of the Noble Gas Family in Group 8A on your Periodic Table. When you see any little element that is purple like Neon, you will know that he is in Group 8A, and a member of the Noble Gas Family. Now you will not have to look at the chart to know the family name of little purple elements. They are the Noble Gases. Eventually you will know all the families' colors like this. Until then you will have my chart to help you identify the Chemical Family an element belongs to using color. The chart will be your friend until then.

Before we leave to meet the families, I'm going to point out the main things you need to know about Chemical Families. After that, it's off to see the Chemical Families and I know you'll love your new adventure.

1. Color and the Chemical Family

"From the first day you met the elements you saw that they were different colors. Now you know why. The color identifies the element's family. See if you can tell me about one element whose color you remember. Check the chart and tell me the family name."

Guy said, "I remember Sodium because we saw him so often. His color was light green." Guy looked at the Chemical Family Chart and light green was the color of the elements in Group 1A . He saw the family name right there—the Alkali Metals."

One more question.-"What's the family name for yellow elements?"

Guy looked at the Chemical Family Color Chart and said, "Yellow elements are in Group 6A and members of the Oxygen Family"

Professor Terry noted——

Guy can use color to find an element's Chemical Family name.

2. Color, Chemical Families and Group

Professor Terry had another question. If at a distance you see a little brown element, can you tell me his Group # as well as his Family Name?"

Looking at the chart, Guy was quick to respond, "Brown is the color of Group 2A elements. He's in Group 2A and he is a member of the Alkaline Earth Metal family."

"If you see a light gray element, tell me his Group # and family name?"

Guy looked at the chart and found light gray color. It was Group 3A on the chart and the element belongs to the Boron family.

THE ELEMENT'S COLOR, GROUP # AND FAMILY NAME

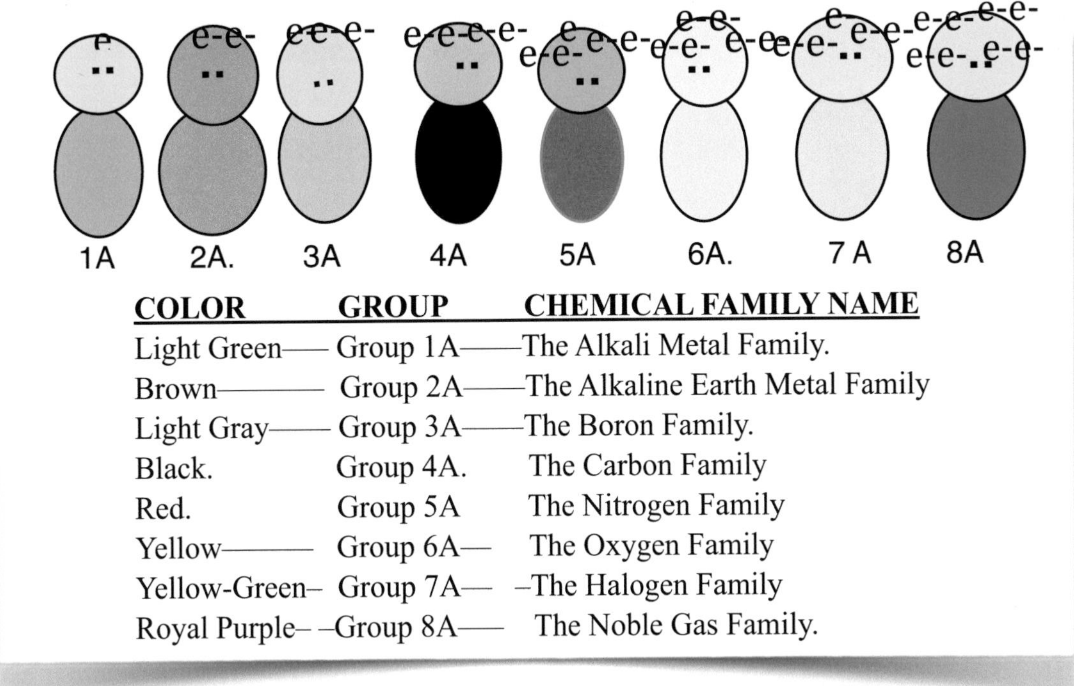

COLOR	GROUP	CHEMICAL FAMILY NAME
Light Green—	Group 1A—	The Alkali Metal Family.
Brown—	Group 2A—	The Alkaline Earth Metal Family
Light Gray—	Group 3A—	The Boron Family.
Black.	Group 4A.	The Carbon Family
Red.	Group 5A	The Nitrogen Family
Yellow—	Group 6A—	The Oxygen Family
Yellow-Green–	Group 7A—	–The Halogen Family
Royal Purple–	–Group 8A—	The Noble Gas Family.

"Here's another chart to show you Color, Group # and Chemical Family Name. Every Group on the Periodic Table is the home of one special Chemical Family. So if you know the Group # you know the Chemical Family. In Periodic Table Land the elements in the same family are all the same color. If you see the color you know the Group# and the Name of the Chemical Family that lives there."

Professor Terry noted—

Guy can use color and the Chemical Family Chart to find the element's Group number as well as his Chemical Family name

3. Color, Chemical Families, Group #, and the Outside Energy Level

"Guy do you remember what the Group # tells us?"

Guy answered, "**The Group # tells us the number of electrons in the Outside Energy Levels of every element in that Group.**"

You know all the elemenets in a Group are in the same Chemical Family. **All the elements in a Chemical Family have the same number of electrons in their Outside Energy Level** because that's what the Group is all about. If you see an atom you can count the number of electrons in its Outside Energy Level, and you will know which Group it is in and which Chemical Family it belongs to as well as its family color." It's all related.

"So Guy, I have an atom here for you. Tell me all about it.

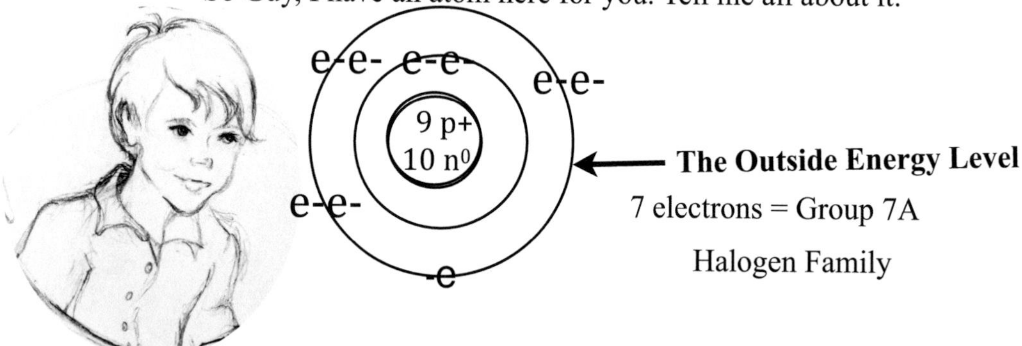

Guy wrote next to the atom—It has 7 electrons in its Outside Energy Level. That puts him in Group 7A on the Periodic Table. Looking at the Chemical Family Color Chart Guy found the Chemical Family name is the Halogen Family and its color is yellow green."

Professor Terry was pleased. So she said, "Here's a chart of the Groups, the Electrons in the Outside Energy Level & Chemical Family Name

Outside Energy Level	Group Number.	Chemical Family
1 electron	Group 1A	Alkali Metals
2 electrons	Group2A.	Alkaline Earth Metals
3 electrons	Group 3A	Boron Family
4 electrons.	Group 4A.	Carbon Family
5 electrons.	Group 5A.	Nitrogen Family
6 electrons	Group 6A	Oxygen Family
7 electrons.	Group7A.	Halogen Family
8 electrons	Group 8A	Noble Gas Family

Professor Terry noted——

Guy knows the elements in a Chemical Family are in the same Group and have the same number of electrons in their Outside Energy Level."

4. How Elements in a Chemical Family Are Different

We just learned that the elements in a Chemical Family are alike because they have the same number of electrons in their Outside Energy Level. But one look tells you these elements that are so much alike are indeed different. Why? Each element in a family is different because it has a different number of energy levels. There are two ways of showing this.

#1 The Period shows the number of energy levels in an atom.

The Period shows us the number of energy levels for all the elements in that row beside the Period Number. On this chart Period 2 begins with Li, and ends with Ne. All these elements have 2 energy levels in their atoms. Elements in Period 3 stretch from Na all theway over to Ar. The 3 before Na says all the elements' atoms have 3 energy levels."

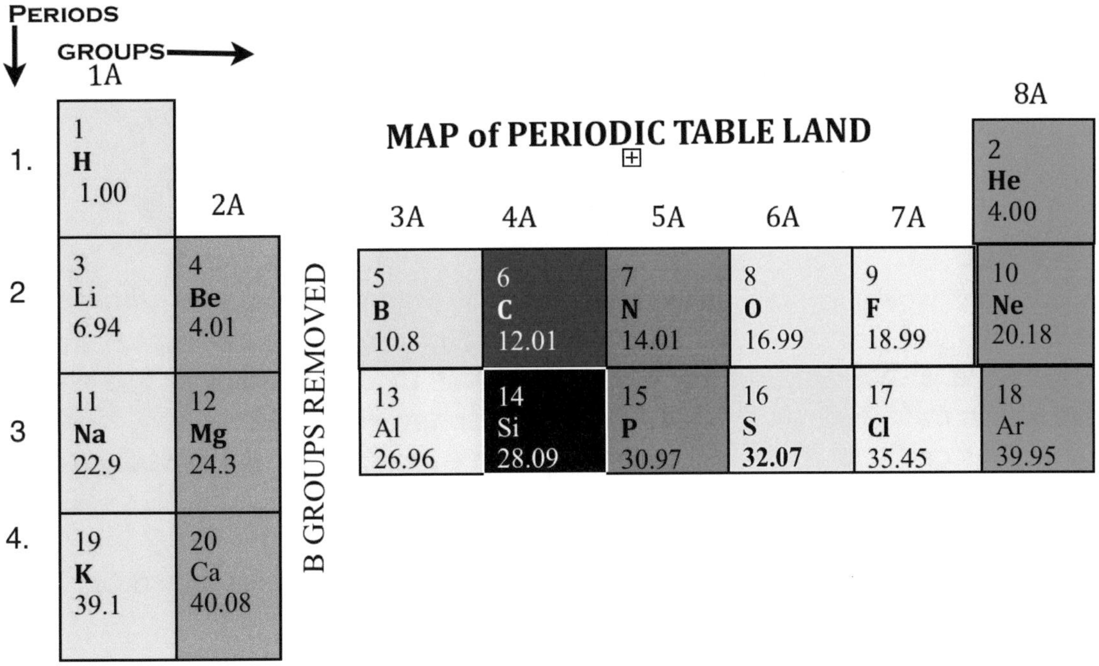

MAP of PERIODIC TABLE LAND

PERIODS ↓ GROUPS →

B GROUPS REMOVED

PERIODS	1A	2A	3A	4A	5A	6A	7A	8A
1.	1 **H** 1.00							2 **He** 4.00
2	3 Li 6.94	4 **Be** 4.01	5 **B** 10.8	6 **C** 12.01	7 **N** 14.01	8 **O** 16.99	9 **F** 18.99	10 **Ne** 20.18
3	11 **Na** 22.9	12 **Mg** 24.3	13 Al 26.96	14 Si 28.09	15 **P** 30.97	16 **S** **32.07**	17 **Cl** 35.45	18 Ar 39.95
4.	19 **K** 39.1	20 Ca 40.08						

"Guy this is not new to you.When you drew Bohr models you drew the number of energy levels equal to the Period Number.

Guy responded, "Yes I remember best Calcium the last Bohr model I drew. I wrote Calcium is in Period 4. So draw 4 energy levels around the nucleus."

"Using the chart above, name the Period for the following elements. I want to make sure you can find the Period for each element: Be, F, K, He, Cl, Mg, O, P, B, Ca, and H. Remember the Period extends all the way across to Group 8A."

Guy said, "In **Period 1** is **H** and **He**. In **Period 2** is **Be, B O and F.** In **Period 3** is: **Cl, P** and **Mg.** In **Period 4** is **K** and **Ca**

#2. **"Over in Periodic Table Land there is a second way to tell how many energy levels an element has. The elements have symbols on their bodies.**

Each element in a Chemical Family has one of these markings on his body. The body marking tells how many energy levels the element has in his atom. Above are the symbols you will see on elements bodies when we visit the Chemical Families. Under the symbol is the number of energy levels it stands for. Just look at the symbol and you can see why it was picked for that number. The only difficult one to see is 7, See if you can locate the 7 parts in the symbol if I tell you the two ends are triangles.

QUIZ

Name the Period and number of Energy Levels for these Element

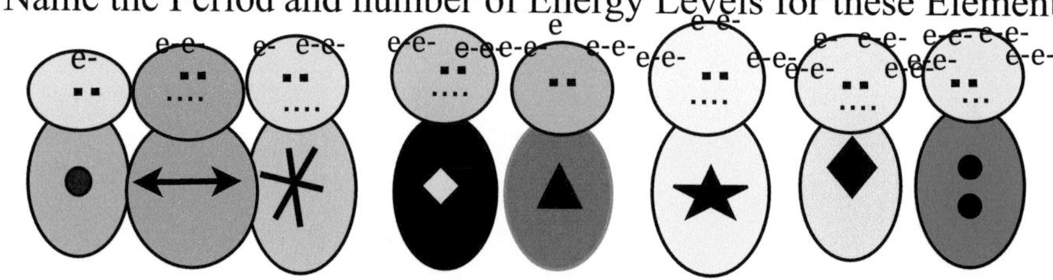

Ans. 1. 7. 6. 4. 3. 5. 4. 2.

Guy wrote only one number under each little element. The number below is both the Period # and the number of Energy Levels in the atom for that element because the Period # and the Number of energy levels is the same. Here's one example. The element with the star on his body is in Period # 5 and has 5 energy levels in his atom.

ProfessorTerry liked how easily Guy learned to interpret the symbols for the Period and Number of Energy Levels.

Professor Terry said, " Suppose you run into this little element in Periodic Table Land, What Period number does the symbol on his body tell you?"

Guy said, "This little elements has a triangle on his body. A triangle has 3 sides. That means he is in Period 3. Since the Period # tells the number of energy levels an element has, this element has 3 energy levels. You didn't ask but he has 1 electron in his Outside Energy Level and he's a light green color. That makes him an Alkali Metal in Group 1A. His one e hair also tells us that he's in Group 1A."

Professor Terry noted——

Guy knows the elements in a family are different because they have different numbers of energy levels. Guy knows that the number of Energy Levels equals the Period Number on the Periodic Table.

5. Locate an Element with Group Numbers and Periods

"Guy, I have just one more skill that I need to know you are able to do before we go to Periodic Table Land. Here it is. I want to make sure you can use the Periodic Table on Page 12 and the Element Name/Atomic Number Chart on Page 84 to find the name of an element using the Group Number and Period."

Professor Terry Has One Final Quiz

THE QUIZ

What is this element's name?

1. What is the Group Number?(count the e hair)_______
2. What is the Period Number? (count the points on the star)______
3. Next go to the **Periodic Table** on Page 12 to find the Group #,_____ and run your finger down to the Period #_______
4. What is the element's symbol. _______?
5. What is the Atomic Number ____ _____?
6. To find the element's name go to Page 84. Look for the Atomic Number on the chart. His name will be next to his Atomic Number. The little element's name is _______

How Guy Solves for the Element's Name

Guy repeated his challenge,"What is this Element's name?" He began by counting the electrons on the element's head–6. He said aloud, "That means the element is in Group 6A. The star on his chest has 5 points. So, the element is in Period 5. First, I have to find Group 6A on the Periodic Table (Page 12) and go down to Period 5. There I will find this element's Atomic Number and symbol." Guy found it and exclaimed, "It is element Atomic Number 52, symbol Te." His face just lit up! He realized how much he had leaned.

Professor Terry reminded Guy to take out the chart listing the elements' names associated with the Atomic Numbers (Page 84). "I think you remember why I gave you this list. Our Periodic Table was too small to write the names of the elements in those tiny boxes. Larger Periodic Tables have the names right in the element boxes." When Professor Terry saw that Guy had found the list in his important papers folder, she was pleased that Guy was developing the good habit of being organized.

Guy looked at the list and found element 52. It was Tellurium symbol Te.

Professor Terry said, "When we visit Group 6A the home of the Oxygen Family, you will get to meet Tellurium"

Guy was thrilled.

Answers to Quiz: 1. Group 6A,. 2. Period 5, 3. Symbol Te, 4. Atomic Number 52
5. Name of Element Tellurium

Professor Terry Noted
Guy Is Ready To Go To Periodic Table Land

Professor Terry thought about what Guy knew about Chemical Families.

1. The Group Number tells the location of a Chemical Family on the Periodic Table.
2. The Group Number tells the number of electrons in the Outside Energy Level.
3. Elements in a family have the same number of electrons in their Outside Energy Level.

4 The Period tells the number of energy levels an element has.

5. Each element in a family is different because they each have a different number of energy levels.
6. If you know the Group # and the Period, you can find the element on the Periodic Table.

After going down her check list Professor Terry was pleased to tell Guy, "You are ready to meet the Chemical Families. We can leave right now for Periodic Table Land."

Guess who appeared? At that very moment Wish Star hopped on to the Lab table. saying, "Hi Professor Terry. I couldn't let you go off on a new adventure without coming to wish Guy well. I keep watching Guy's progress from above. He's learning so much. You are a great teacher Professor Terry. I'm so glad I brought Guy to you when I couldn't stay for the whole summer. Guy's wish is coming true. I see he's learning so much about the universe below the magnificent world he loves. I've got to go now but enjoy the Chemical Families". In an instant he was gone as magically as he arrived.

When Wish Star left Professor Terry said, "Are you ready Guy? "
Guy said, "I'm always excited to go. It's my favorite place."
They headed to the back of the lab and the magic entrance to Periodic Table Land.

THE ALKALI METAL FAMILY

Chapter 2

In the back of the lab Professor Terry and Guy, leaned on The Periodic Table. The whole section of wall pushed in and flipped around. Bells, whistles and sirens screamed—adventure ahead. Professor Terry said, "Here we go!" As she held Guy's hand they found themselves sliding down a tunnel of mirrors reflecting a thousand sparkling lights. Magic sprinkle dust enveloped them like a silky blanket. They felt comforted by soft mysterious music. Then, in an instant, they were in that strange new world that fascinated Guy so much. Guy was still mesmerized no matter how many times he went there. Being greeted by those festive colorful blinking lights surrounding the street signs forever intrigued him. The street signs were once the Group Numbers on the Periodic Table. Now Guy knew that the color of the street lights was the Chemical Family's special color.

Professor Terry said, "I know you'll enjoy meeting the Chemical Families. Many of these elements you have already met while drawing their Bohr models. But there are many that you have not. You will enjoy visiting once again with the elements you already know, and you will love meeting the elements you have never met before. It's important to get to know all these elements because you will be running into them again as you learn how compounds are formed and all about chemical reactions. We are going to begin on 1st Street, and go from there till we meet all the families ending on 8th Street."

Guy was familiar with 1st Street as he had been there so many times before visiting Sodium. He also had drawn Bohr models for Hydrogen, Lithium, Sodium and Potassium. He really had met most of the elements on 1st Street but there were still some he had not met..

"Look Guy 1st Street now has a flag. This is something new. They were near the entrance to 1st Street when they saw the flag."

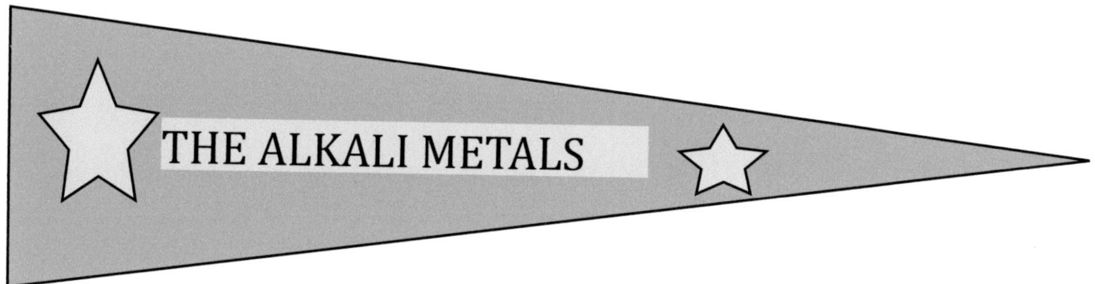

The flag was rather beautiful Guy thought as they walked down 1st Street, Group 1A on the Periodic Table, the home of the Alkali Metal family. Guy was excited as he realized their adventure was truly starting. As they made their way down the street, Guy noticed the street was painted green–the same color as their flag which was flapping slightly in the light breeze.

Professor Terry shared, “You will soon see that the members of this Alkali Metal Family are all going to be green too. It is the official color of their Chemical Family. the Alkali Metal Family.There’s Sodium’s place up ahead. We are almost there.”

Guy Meets the Alkaline Earth Metal Family
1st Street in Periodic Table Land

PERIODS ↓ **GROUPS** →

THE PERIODIC TABLE OF THE ELEMENTS

	1A	2A	3B	4B	5B	6B	7B	8B	8B	8B	1B	2B	3A	4A	5A	6A	7A	8A
1.	1 H 1.00																	2 He 4.00
2.	3 Li 6.94	4 Be 4.01											5 B 10.8	6 C 12.0	7 N 14.0	8 O 16.9	9 F 18.9	10 Ne 20.1
3.	11 Na 22.9	12 Mg 24.3											13 Al 26.9	14 Si 28.0	15 P 30.9	16 S 32.0	17 Cl 35.5	18 Ar 39.9
4	19 K 39.1	20 Ca 40.0	21 Sc 44.0	22 Ti 47.9	23 V 50.9	24 Cr 51.9	25 Mn 54.9	26 Fe 56	27 Co 58.9	28 Ni 58.6	29 Cu 63.5	30 Zn 65.3	31 Ga 69.2	32 Ge 72.6	33 As 74.9	34 Se 78.9	35 Br 79.9	36 Kr 83.7
5.	37 Rb 85.5	38 Sr 87.6	39 Y 88.9	40 Zr 91.2	41 Nb 92.9	42 Mo 95.9	43 Tc {98}	44 Ru 101	45 Rh 103	46 Pd 106	47 Ag 107	48 Cd 112	49 In 114	50 Sn 119	51 Sb 122	52 Te 126	53 I 126	54 Xe 131
6.	55 Cs 132	56 Ba 137.	57 - 71	72 Hf 178	73 Ta 180	74 W 183	75 Re 186	76 Os 190	77 Ir 192	78 Pt 195	79 Au 196	80 Hg 200	81 Tl 204	82 Pb 207	83 Bi 208	84 Po 209	85 At 210	86 Rn 222
7	87 Fr 223	88 Ra 226	89-1 03	104 Rf 267	105 Db 268	106 Sg 271	107 Bh 272	108 Hs 270	109 Mt 278	110 Gs 281	111 Rg 280	112 Cn 285	113 Nh 284	114 Fl 289	115 Mc 288	116 Lv 203	117 Ts 294	118 Og 294

57 La 139	58 Ce 140	59 Pr 141	60 Nd 144	61 Pm 145	62 Sm 150.	63 Eu 151	64 Gd 157	65 Tb 159	66 Dy 163	67 Ho 165	68 Er 167	69 Tm 169	70 Yb 173	71 Lu 175
89 Ac 227	90 Th 232	91 Pa 231	92 U 238.	93 Np 237	94 Pu 234	95 Am 243	96 Cm 247	97 Bk 247	98 Cf 251	99 Es 252	100 Fm 257	101 Md 258	102 No 259	103 Lr 262

When they arrived at #11, 1st Street, Sodium popped out from behind his house and greeted them, “Well, isn’t this a pleasant surprise! So good to see you. Have you come this way for a reason? I bet you learned about our Flag Festival!”

“I noticed your new flag, but I didn’t know a festival was going on. This is a good time to visit. Guy wants to learn all about your Chemical Family,” responded Professor Terry.“Actually, I plan to take Guy all across Periodic Table Land to meet each of the Chemical Families. I want him to get to know the as many elements as possible.”

Sodium responded, “I’ll be happy to introduce you to the elements in my family. When you visit each element, you’ll see them working on a festival patch for our family banner that we’ll hang alongside our flag. All the Chemical Families are entering their flags and banners in a contest hoping to win a prize. The squares for the banners will display the element's favorite use or property. A property is physical when it describes the element; it's chemical if it tell what the element does.. I made a small patch for the top of our banner. Let me show you."

The Alkali Metals

"Professor Terry, I heard you explaining to Guy about our color, and you were correct. Everything about us is green because it's our family color. We're the Alkali Metals, and our color distinguishes us from elements in the other Chemical Families. You will also notice each member of our family has one electron on our head. This is because we are all in Group 1A. Guy, I think you'll remember from drawing our Bohr models, that members of our family also have only one electron in our Outside Energy Level. This is the family characteristic that makes us so much alike. We are not like the usual, hard metals that are used to make things. We are shiny like metals, but we are very light weight and rather soft.

"Most of our family members are extremely reactive. That means we are always ready to combine with other elements. Some of us even explode when put into water. That can be dangerous. The more energy levels an element has, the more reactive it is. Elements in Period 6 with 6 energy levels are more reactive than the elements in Period 2 with two energy levels. Come now. The elements are anxious to meet you and to show you the patches they have made for our banner.They are all excited because of the flag festival. We'll visit Lithium first."

LITHIUM

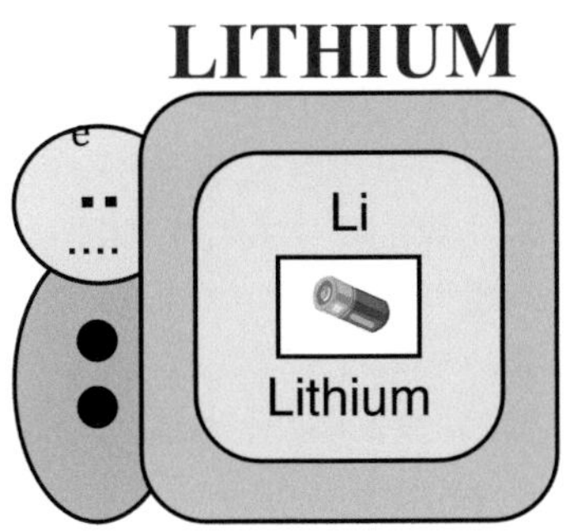

LITHIUM was sitting on his porch, working on the square he was making. "I'm the lightest metal in our family," said Lithium. "Remember how small my atom looked when you drew my Bohr model? I have just 2 energy levels. I am a soft silvery metal and I can make a flame turn red. I told you how I'm used in drugs for patients with a bipolar disorder. I'm also used to make Lithium batteries for cell phones and watches. For my square I'm drawing a battery."

SODIUM

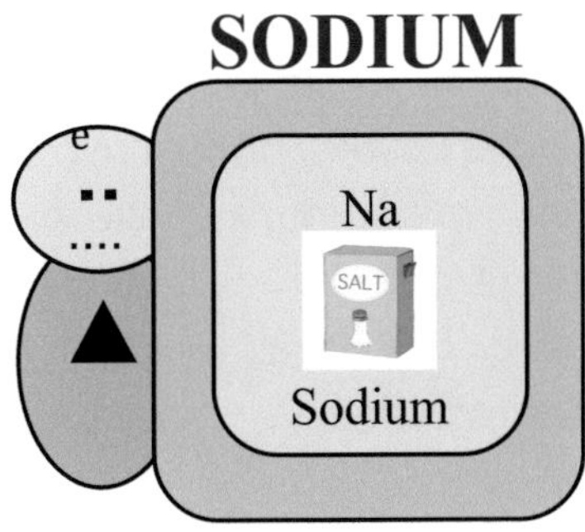

SODIUM said, "Notice my body's triangle has 3 sides meaning I'm in Period 3. I'm sure you remember that you gave me 3 energy levels when you drew my Bohr model. I control people's blood pressure and the amount of fluid in the body. When I form a compound with Chlorine, I make a beautiful white crystal, salt. I enjoy making your food taste so good. I'm found in sea water which covers 3/4 of our earth. People call me 'Old Salty' for short. Here's the patch I've made for our banner. I drew a box of salt."

POTASSIUM

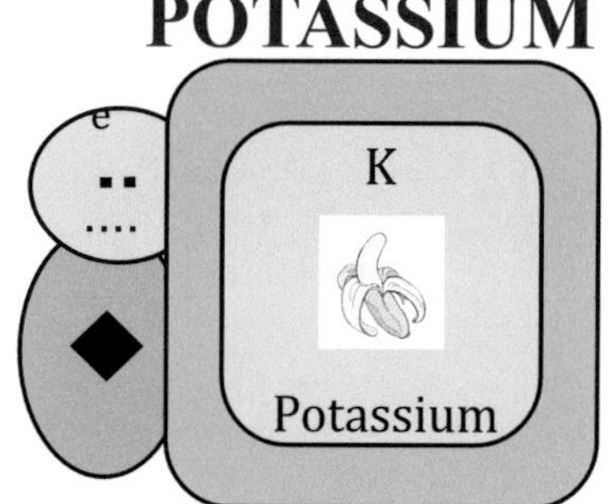

POTASSIUM was next on the list to visit. "Did you notice the four sided diamond on my body? I'm in Period 4, and I have 4 energy levels in my atom." I'm found in blood. There, I work with Sodium to regulate the heart and the kidneys. I give a lilac (pink) color to flames. The pink flame is used to test for me. Some of my compounds are fertilizers that farmers use to grow better crops for you to eat. Bananas have lots of Potassium in them. I'm drawing a banana on my square."

RUBIDIUM

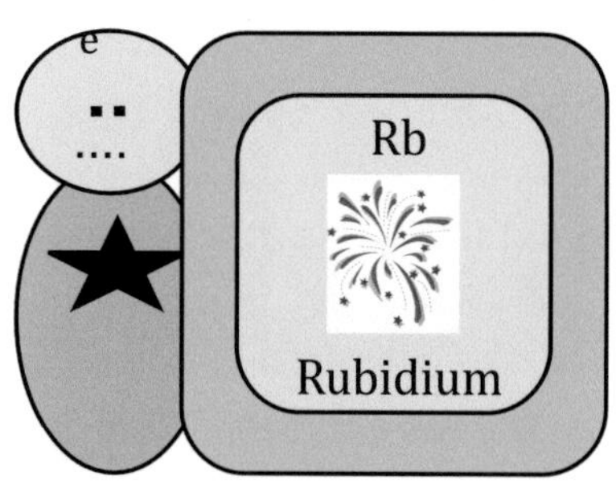

RUBIDIUM saw them and walked over to speak with Sodium and Guy. "Hi Guy! I've heard a lot about you, but we have never met. As you can see I'm famous for some of the beautiful fireworks that you enjoy on the Fourth of July." Guy noticed the 5 pointed star on his body. So he knew he was in Period 5 and had 5 energy levels. "I'm getting ready for the 4th of July celebration. I'm practicing my fireworks. That's what I'll draw on my square. Don't you love the red color of my fireworks?"

CESIUM

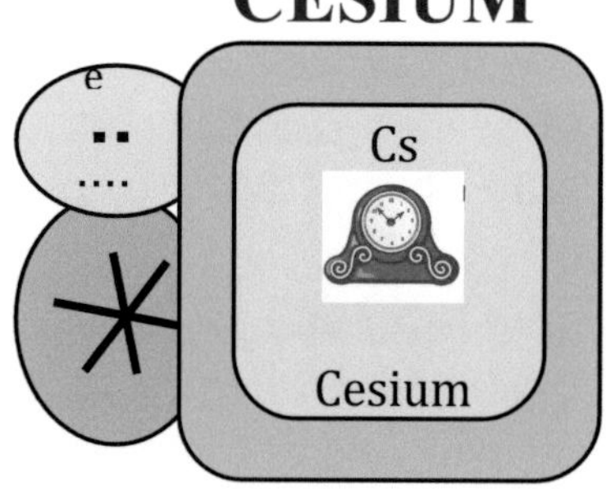

CESIUM said, "Hi Guy! I give flames a blue color. That's a color test they use to identify me."

Guy said, "I notice you have a lot of energy levels. You must be more reactive than all the elements I met so far in your family." Cesium agreed. Then he said, "An atomic clock was built a while ago using me. A second on that clock was the most precise measurement of a second in time. To this day Cesium remains the standard for a second in time." Cesium told Guy, "I am most proud of this. So I drew a clock on my patch."

FRANCIUM

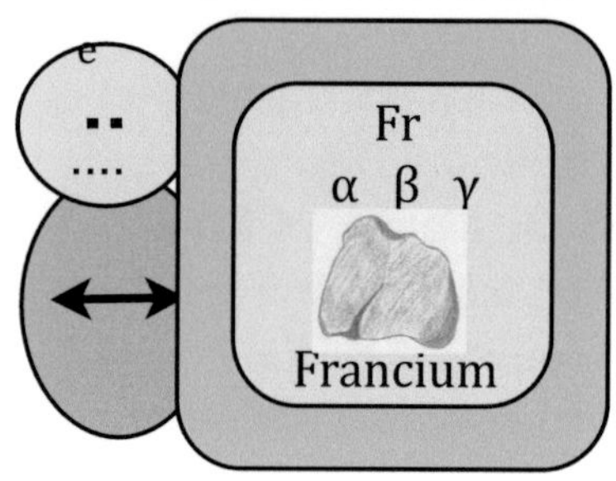

FRANCIUM. "I'm the heaviest of all the Alkali Metals. Francium explained that this double sided arrow symbol on his chest meant he had 7 energy levels in his atom. Then Francium added, "I'm also radioactive. I drew a rock with radiation coming out of it on my square. The Greek letters alpha, beta and gamma are symbols for radiation.

"Now that you've met all the members of my family, I'm sure you're ready to move on."

Guy said, "I have a question for you Sodium. Why did you skip over Hydrogen, isn't he in Group 1A?"

"You are right, Guy, he is in Group 1A, because he has 1 electron in his Outside Energy Level. There's no question that he needs to be in Group 1A. However he's a gas and can't possibly be called a metal—an **Alkali Metal**."

Before Guy had a chance to leave, all the Alkali Metals came running over to Guy and presented him with a card saying, "This contains the important information about our family. The diagram of our atoms' Outside Energy Level shows one electron there. That electron will become most important when you learn how we form compounds. For now just remember that all the Alkali Metals in Group 1A have one electron in their Outside Energy Level."

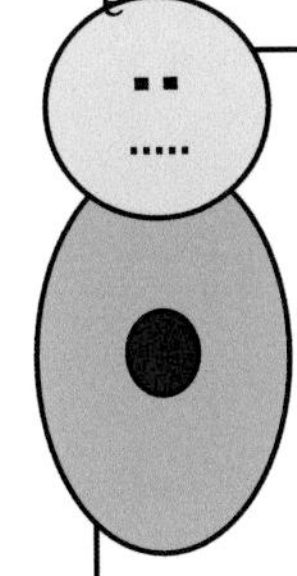

CHEMICAL FAMILIES
GROUP 1A
THE ALKALI METALS

In Group 1A, each element has 1 electron in its Outside Energy Level

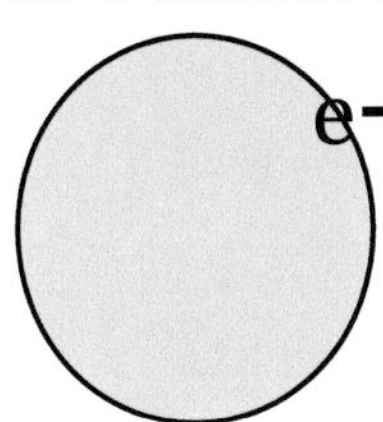

3 Lithium	(Li) in medicine used to treat people with bipolar disease.
11 Sodium	(Na) in salt that we use to season food.
19 Potassium	(K) with Sodium it makes our heart pump properly.
37 Rubidium	(Rb) gives red color to fireworks.
55 Cesium	(Cs) is the International definition of a 'second' of time.
87 Francium	(Fr) is radioactive and the heaviest atom in the family.

Sodium and Guy returned to find Professor Terry waiting on the front porch. . Sodium said, "I expect that you are going to need a way to get around Periodic Table Land. Our magic bubble is yours to use."

Professor Terry thanked Sodium. Guy told how much he enjoyed meeting all the elements. Then they climbed into the magic bubble and off they went. When they were high enough they could see the excitement the festival had brought to Periodic Table Land. Elements were running around. All thc colorful flags brought color to the Land. It was a great time to visit the Chemical Families.

THE ALKALINE EARTH METAL FAMILY
Chapter 3

2nd Street was right below now. Professor Terry pointed out to Guy, "Notice 2nd Street is painted brown to get ready for the celebration. It's the official color of the Alkaline Earth Metals." Guy pointed the magic wand and they soon landed.

At the entrance to 2nd Street, Group 2A on the Periodic Table, there was a large earth tone flag announcing the name of the Chemical Family that lived there. In golden letters it read The Alkaline Earth Metals.

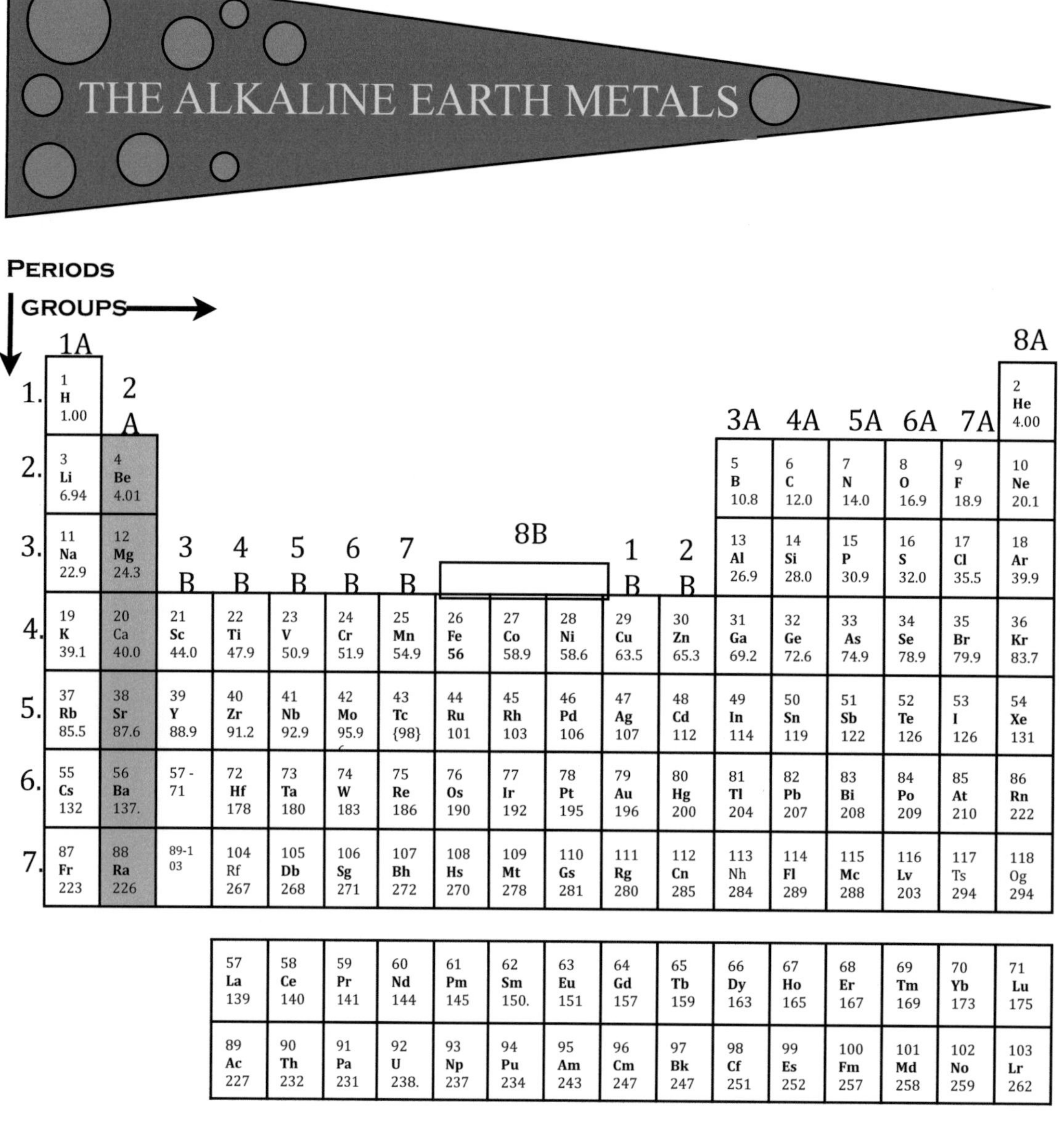

	1A	2A	3B	4B	5B	6B	7B	8B			1B	2B	3A	4A	5A	6A	7A	8A
1.	1 H 1.00																	2 He 4.00
2.	3 Li 6.94	4 Be 4.01											5 B 10.8	6 C 12.0	7 N 14.0	8 O 16.9	9 F 18.9	10 Ne 20.1
3.	11 Na 22.9	12 Mg 24.3											13 Al 26.9	14 Si 28.0	15 P 30.9	16 S 32.0	17 Cl 35.5	18 Ar 39.9
4.	19 K 39.1	20 Ca 40.0	21 Sc 44.0	22 Ti 47.9	23 V 50.9	24 Cr 51.9	25 Mn 54.9	26 Fe 56	27 Co 58.9	28 Ni 58.6	29 Cu 63.5	30 Zn 65.3	31 Ga 69.2	32 Ge 72.6	33 As 74.9	34 Se 78.9	35 Br 79.9	36 Kr 83.7
5.	37 Rb 85.5	38 Sr 87.6	39 Y 88.9	40 Zr 91.2	41 Nb 92.9	42 Mo 95.9	43 Tc {98}	44 Ru 101	45 Rh 103	46 Pd 106	47 Ag 107	48 Cd 112	49 In 114	50 Sn 119	51 Sb 122	52 Te 126	53 I 126	54 Xe 131
6.	55 Cs 132	56 Ba 137.	57 - 71	72 Hf 178	73 Ta 180	74 W 183	75 Re 186	76 Os 190	77 Ir 192	78 Pt 195	79 Au 196	80 Hg 200	81 Tl 204	82 Pb 207	83 Bi 208	84 Po 209	85 At 210	86 Rn 222
7.	87 Fr 223	88 Ra 226	89-1 03	104 Rf 267	105 Db 268	106 Sg 271	107 Bh 272	108 Hs 270	109 Mt 278	110 Gs 281	111 Rg 280	112 Cn 285	113 Nh 284	114 Fl 289	115 Mc 288	116 Lv 203	117 Ts 294	118 Og 294

57 La 139	58 Ce 140	59 Pr 141	60 Nd 144	61 Pm 145	62 Sm 150.	63 Eu 151	64 Gd 157	65 Tb 159	66 Dy 163	67 Ho 165	68 Er 167	69 Tm 169	70 Yb 173	71 Lu 175
89 Ac 227	90 Th 232	91 Pa 231	92 U 238.	93 Np 237	94 Pu 234	95 Am 243	96 Cm 247	97 Bk 247	98 Cf 251	99 Es 252	100 Fm 257	101 Md 258	102 No 259	103 Lr 262

Guy Meets the Alkaline Earth Metal Family 2nd Street in Periodic Table Land

With some good fortune, Calcium was standing right by the 2nd Street gate as they approached. Professor Terry and Guy were good friends with Calcium dating back to Guy's adventures with Calcium's fun loving electrons. Professor Terry was glad to see her dear friend.

Calcium said, "How do you like our family flag? We're rather proud of it as it represents our family. Everyone treats it with so much respect because it represents the family we love."

"It's beautiful and rather appropriate for your earthy family," said Professor Terry.

Guy said, "I like it because those little globes on it remind me of the planets. I love the night sky. As I sit watching the stars, I imagine the planets whirling around out there in space. Your flag brings back happy memories."

Calcium started to tell Guy about his family. "We're metals and very reactive. That means we like to combine with other elements whenever we can find them. We're like the Alkali Metals, but they are even more reactive. We are principally noted for creating alkaline compounds. That's how we got our name. Alkaline compounds are the direct opposite of acid compounds which most people have heard about. Both acids and bases are strong chemicals."

"You arrived at a good time. You know there's a contest going on, and each of the Alkaline Earth Metals is making a little square showing everyone what his element is like. I made this little rectangle to go at the top of our family banner."

"Let's start your visit with Beryllium. You drew his Bohr model so it will be fun seeing him again." They strolled slowly down 2nd Street talking about the fun Guy had with Calcium's electrons. Before they knew it they were at the home of Beryllium, Atomic Number 4. That made his address #4, 2nd Street.

BERYLLIUM

BERYLLIUM pointed to his banner patch and said, "I'm used in many kinds of aircraft as well as space craft. So I'm drawing an airplane on my patch. I am found in the emerald, a precious green gem. I form an alloy with copper which makes me wear resistant and useful for gyroscopes. I create alloys with nickel, and I'm used in spark resistant tools."

“Well Guy, the first member of our family might be small, but his uses are quite varied. What did you like best?”

Guy said, “That’s easy. If I were Beryllium, I would enjoy being part of that beautiful precious green gem, the emerald.”

“Let’s move on to visit the next element in my family——Magnesium.” They chatted as they walked. Then Guy said, “What I remember best was when your little electrons sang their welcome to me. Then I loved when they grabbed hold of my hand and led me up on to the roof of your house showing me the Periods way over on the edge of town.” They chatted a little longer and soon arrived at #12, 2nd Street, the home of Magnesium

MAGNESIUM

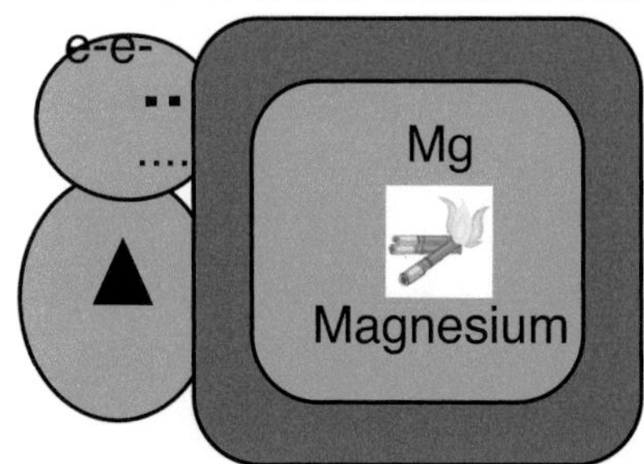

MAGNESIUM was just coming out of his house as they arrived. Magnesium said, “It was hard to decide what to draw for my patch because I’m found in abundance in the Earth’s crust, in the stars, and in the ocean. I’m also an important element in the human body and in plants. I decided to draw marine flares on my square because anyone lost at sea can use these flares to help searchers find them.” Magnesium’s compounds are called alkaline and they neutralize acids. “Everyone loves my bright white fireworks.” Guy thanked Magnesium and moved on to see Calcium.

As they moved along 2nd Street Guy continued talking with Calcium, “I also remember the funny story the electrons told me about their negative charges.” Just then Calcium pointed to his own house.

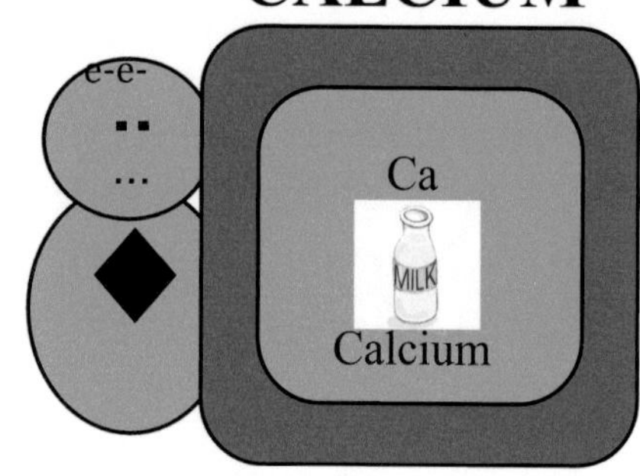

CALCIUM said, “My house is next. You have known me for a long time, but did you know I’m found in bones and teeth? I couldn’t decide if I should draw bones or a tooth on my patch because I make both of these parts of the body strong. So instead, I decided to draw a bottle of milk because milk is such a good source of Calcium. I want to remind people to drink milk to keep their bones and teeth strong.”

They chatted a little longer, and the house that they came to next was the home of Strontium. He was out in his garden resting in his hammock looking so comfortable. They walked over to Strontium and greeted him. Strontium said, “It’s summer. So I like being out in my garden as much as possible. Then he told Guy all about himself.

STRONTIUM "I heard you met my friend element, Rubidium. I make fire works red like he does. We're old buddies and enjoy the 4th of July. I once was used in making TVs, but I've been replaced. I'm also used to make red glass and to improve pottery glazes. I'm drawing a picture of a red crystal cup on my patch."

BARIUM spoke up. "If you have a problem called reflux, the doctor schedules you for a Barium swallow. This very different compound of mine makes your esophagus show up on an x-ray. That lets the doctors see what's wrong. I have compounds that are used to keep down the rat population and to keep houses free of insects. For fun, I make fireworks. I'm also found in white paint. I'm drawing a can of white paint on my square."

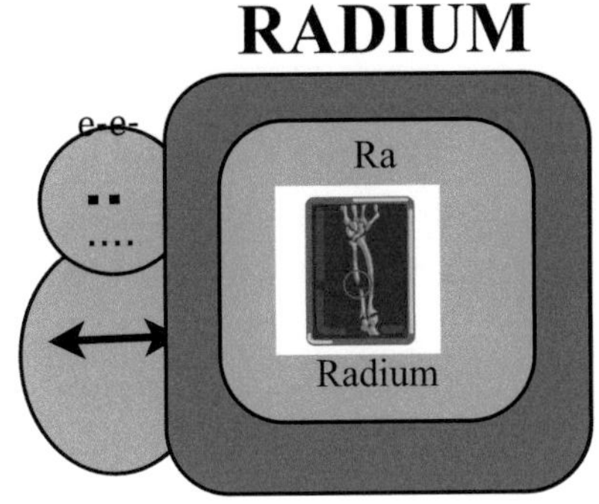

RADIUM whispered, "If you have ever broken bones in your body, your doctor has put me to good use. Since I'm radioactive, I'm a source of X-rays. This helps the doctor see if your bones are broken. I also help to see if you have lung cancer or pneumonia. I'm putting on my square an X ray of a broken bone."

The little elements in the Alkaline Earth Metal family ran over and gave Guy a card.

CHEMICAL FAMILIES
GROUP 2A
ALKALINE EARTH METALS

Alkaline Earth Metals are light weight metals and very reactive. They are not as active as the Alkali Metals. Their most notable chemical property is that they form alkaline compounds. This is a strong chemical, the opposite of an acid.

4 **Beryllium** (Be) is used in high performance aircraft and spacecraft.
12 **Magnesium** Mg) is used in medicine for acid stomachs.
20 **Calcium** (Ca) is found in bones and teeth.
38 **Strontium (**Sr) is used in making fireworks.
56 **Barium** (Ba) is used to make the digestive tract show up on X-ray film.
88 **Radium** (Ra) gives off x-rays and is used to make things glow in the dark.

They told Guy, "The most important thing to remember is that we have two electrons in our Outside Energy Level. It is these electrons that are involved in creating compounds. It's easy to remember. If the Group number is 2A, there are 2 electrons in the Outside Energy Level. That's what the Group number means."

Guy took another look at the Alkaline Earth Metals' chart. He paid special attention to the little brown circle which represents their Outside Energy Level with the two electrons. He said, "I'll remember that means these elements are in Group 2A. The Group 2A will now have a new meaning. Group 2A will make me think of this wonderful Alkaline Earth Metal Family."

Guy's encounter with Calcium before had been mainly with Calcium's delightful electrons. Now he had met all the elements in the Alkaline Earth Metal Family. That was so good. Guy was told that more of these elements formed alkaline compounds than any other family. Alkaline is the opposite of acid. Professor Terry told Guy that he would understand more about acids and alkaline chemicals later when he studied different kinds of compounds."

. Then, Professor Terry and Guy slid into the bubble, and continued their journey.. The view of the lovely countryside below made Guy recall that learning more about his magnificent world was the reason he wanted to study chemistry. Guy exclaimed, "When I drew Bohr models I didn't get to meet the Transition Metals. I can't wait to visit the elements that live in those interesting house on B Avenue that we flew over so often. Off they flew to visit the Transition Metals.

THE TRANSITION METAL FAMILY
Chapter 4

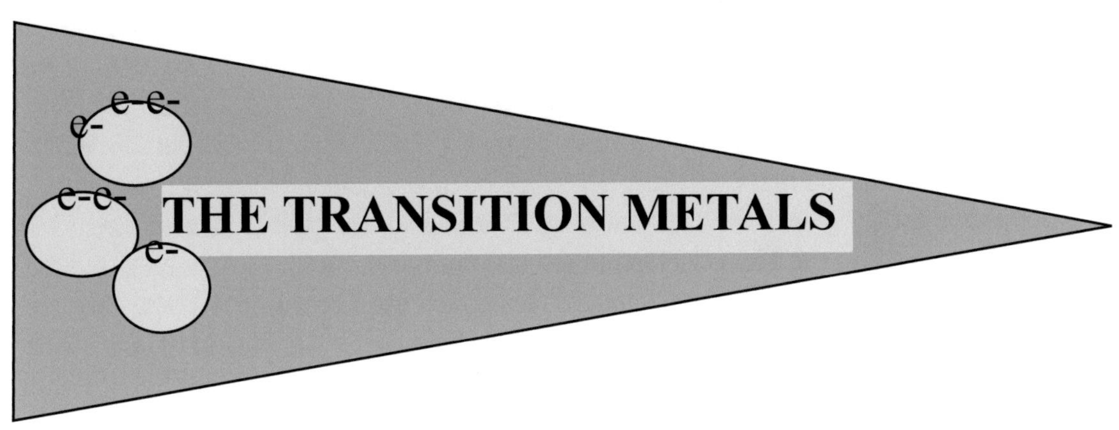

PERIODS

GROUPS →

THE PERIODIC TABLE OF THE ELEMENTS

Period	1A	2A	3B	4B	5B	6B	7B	8B	8B	8B	1B	2B	3A	4A	5A	6A	7A	8A
1.	1 H 1.00																	2 He 4.00
2.	3 Li 6.94	4 Be 4.01											5 B 10.8	6 C 12.0	7 N 14.0	8 O 16.9	9 F 18.9	10 Ne 20.1
3.	11 Na 22.9	12 Mg 24.3											13 Al 26.9	14 Si 28.0	15 P 30.9	16 S 32.0	17 Cl 35.5	18 Ar 39.9
4.	19 K 39.1	20 Ca 40.0	21 Sc 44.0	22 Ti 47.9	23 V 50.9	24 Cr 51.9	25 Mn 54.9	26 Fe 56	27 Co 58.9	28 Ni 58.6	29 Cu 63.5	30 Zn 65.3	31 Ga 69.2	32 Ge 72.6	33 As 74.9	34 Se 78.9	35 Br 79.9	36 Kr 83.7
5.	37 Rb 85.5	38 Sr 87.6	39 Y 88.9	40 Zr 91.2	41 Nb 92.9	42 Mo 95.9	43 Tc {98}	44 Ru 101	45 Rh 103	46 Pd 106	47 Ag 107	48 Cd 112	49 In 114	50 Sn 119	51 Sb 122	52 Te 126	53 I 126	54 Xe 131
6.	55 Cs 132	56 Ba 137.	57 - 71	72 Hf 178	73 Ta 180	74 W 183	75 Re 186	76 Os 190	77 Ir 192	78 Pt 195	79 Au 196	80 Hg 200	81 Tl 204	82 Pb 207	83 Bi 208	84 Po 209	85 At 210	86 Rn 222
7.	87 Fr 223	88 Ra 226	89-1 03	104 Rf 267	105 Db 268	106 Sg 271	107 Bh 272	108 Hs 270	109 Mt 278	110 Gs 281	111 Rg 280	112 Cn 285	113 Nh 284	114 Fl 289	115 Mc 288	116 Lv 203	117 Ts 294	118 Og 294

57 La 139	58 Ce 140	59 Pr 141	60 Nd 144	61 Pm 145	62 Sm 150.	63 Eu 151	64 Gd 157	65 Tb 159	66 Dy 163	67 Ho 165	68 Er 167	69 Tm 169	70 Yb 173	71 Lu 175
89 Ac 227	90 Th 232	91 Pa 231	92 U 238.	93 Np 237	94 Pu 234	95 Am 243	96 Cm 247	97 Bk 247	98 Cf 251	99 Es 252	100 Fm 257	101 Md 258	102 No 259	103 Lr 262

B Avenue in Periodic Table Land
Guy Meets the Transition Metals

Professor Terry said, "Guy, you're going to meet a few elements whose houses you admired as we flew over B Avenue when you were drawing Bohr models. Remember the house that looked like a frying pan? That's Iron's place, and you will meet him first. Then there was Silver with his sparkling pointed towers. I also remember you liked Gold's golden dome. I also want you to meet copper. I don't think we noticed his house. When we get there maybe there will be time to see copper's house, but maybe not."

Suddenly, Guy gave a twist to the magic wand and the bubble floated down just above the trees and houses. They decided to fly low today over wide B Avenue to capture a better view of the uniquely different metallic homes of the Transition Metals. Guy was amazed that these houses were each so different even though they appeared to be made of the same silvery metal. Their artistic designs were amazing. It was Professor Terry's hope to have Guy meet just a few of the Transition Metals, as there was no time to meet them all. Iron was first on the list. Soon they were there, and that flag was now close enough to clearly read the words Transition Metals. Guy noticed the whole Avenue was painted a dark gray, the family color.

Professor Terry took out her powerful magnet, and the force of magnetic attraction pulled them to Iron's house in a nano second. "We'll set the bubble down by his house, and I'll ask Iron to send for a few Transition Metals. Then you'll finally meet them."

. Iron heard them land and came out to greet them. "Hi there, it's good to see you. What you doing here on B Avenue?"

Professor Terry introduced Guy to Iron and told him the they had come to learn all about the Transition Metals."

Iron said, "Let me tell you a little about myself first."

IRON said, "My symbol is Fe from the Latin word for Iron, Ferrous. I am strong and abundant in the Earth." Iron showed Guy his square. "I am a good conductor of heat. That means I'm good for cooking.So I drew a frying pan on my square I rust easily so they combined me with different elements to make an alloy called steel. Steel is used in construction to make tall buildings. In the body, I am an important part of our blood's hemoglobin which carries oxygen to all parts of the body. I have a lot of uses."

Professor Terry said, "It's wonderful to learn all these facts about you. Now I'd like to hear about your family. But before you start, I have to ask you a favor. Could you please send for several of your Transition Metals to visit with Guy while we are here?"

Iron texted an invitation to several Transition elements to meet Guy and Professor Terry. "While we are waiting," explained Iron, "I'll tell you a little about my Transition Metal family. We are metals as you ordinarily think of metals—hard, shiny metallic solids. However, our elements are not like the elements in the A Group families that have the same number of electrons in their Outside Energy Levels as their Group number indicates. The elements in the Transmission Metal Family don't follow this rule. Take me, for instance. Sometimes I have two electrons on my Outside Energy Level, and then the compounds I form are called ferrous. Sometimes I have three electrons there, and my compounds are called ferric. The design on our flag shows how varied our Outside Energy Levels can be. We even have a metal that is a liquid, Mercury. That's what I mean when I said we are complex. Guy, later you will study us and our rules separately. However, I heard you are just getting started learning the basics of chemistry. When you learn the basics well, understanding our family will be so much easier. I've invited a few of the better known metals in my family to meet you. Silver, Gold, Mercury, Zinc, and Copper have agreed to come. You'll be happy to have met them when you meet them again when you learn about chemical reactions and balancing equations."

."Guy, I see Silver and Gold approaching. Looks like they're the first to arrive."

SILVER AND GOLD showed up at the same time. "We are friends," they said together, "We can both be used for jewelry. We are worth a lot of money."

SILVER AND GOLD

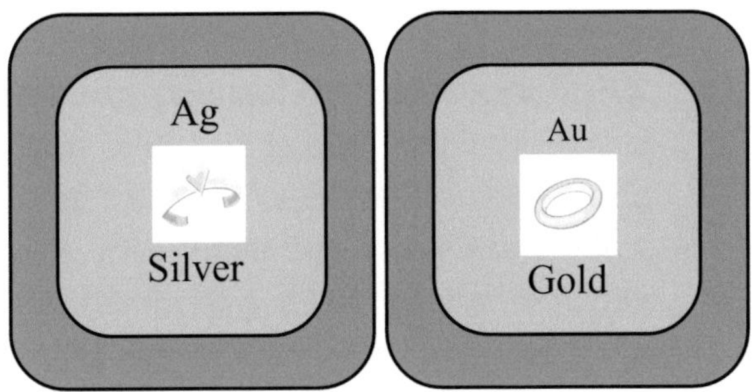

GOLD said, "I'm worth a lot more than silver. In world money circles, I've been called *The Gold Standard.* Both of us are used in coins by many different countries. Both of us are crafted into beautiful pieces of jewelry like rings, necklaces and bracelets. I'm going to draw a man's wedding ring for my patch"

SILVER said, "That's right. I enjoy being made into good jewelry. I also have a special use in photography and medical X-rays because silver based film is sensitive to light and can record a picture. In industry, using x-ray, I can show cracks in metal which they can repair before a structure collapses. I'll draw a bracelet for my festival patch."

COPPER

Suddenly, **COPPER** showed up. "My best use is in conducting electricity because I'm ductile. This means I can be easily pulled into wire. I'm really useful in electrical circles. I am also used by artists to make beautiful sculptures. I was used in plumbing, but now I cost too much and have been replaced by less expensive materials. I can also conduct heat so there are some pots that have copper bottoms. I'm also used in making coins by many nations. For my square, I'm drawing a copper coin."

MERCURY

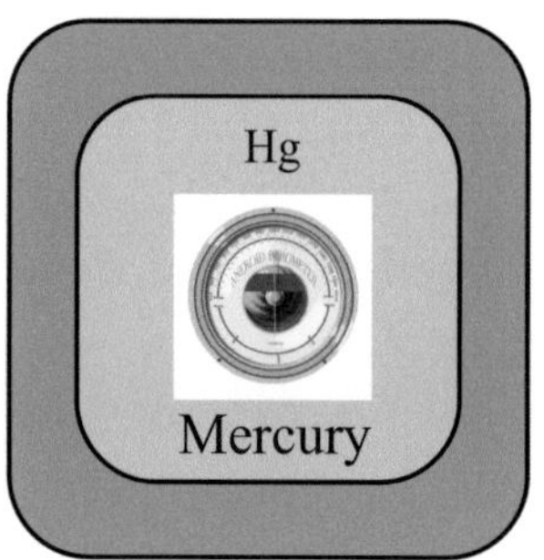

MERCURY. "I heard you wanted to see me, Guy. I guess I'm the only metal in the world that is a liquid at room temperature. I've been around for thousands of years. When I combine with other elements like gold and silver, they call the product I form, amalgams. I had many uses before scientists found out I was poisonous. If I get into a human body, the kidneys can not filter me out. Not too long ago I was that silvery metal in thermometers that expanded with heat to tell people's temperature. I'm drawing a picture of an aneroid barometer for my square A barometer measures air pressure, and it is used to predict the weather.

ZINC

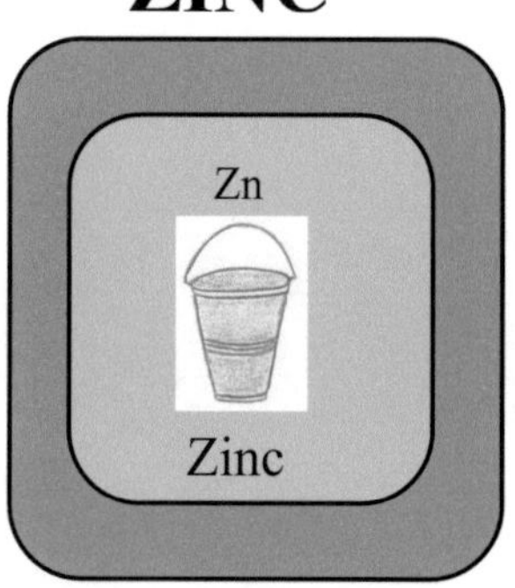

ZINC arrived a little late and said, "You probably haven't heard about me before; however, in chemistry you will be hearing a lot about me. I like to form alloys with copper to make brass which has many uses. In industry, I'm used to galvanize metals that corrode like iron. I cover these metals with zinc, and they last so much longer. My favorite use is to cover garden buckets so that they will not rust. That's what I'm drawing for my patch."

Guy and Professor Terry thanked the Transition Metals for a delightful visit and a chance to learn so many of the symbols based on foreign languages.

Guy thought, "These elements have a few symbols that I don't know. I think I can learn them if I try without too much trouble." Guy read them out loud.

"Cu–Copper, **Fe**–Iron, **Ag**–Silver, **Au**–Gold, **Zn**–Zinc, and for Mercury–**Hg**. "

Guy recalled a story Professor Terry had told him once about how to remember the symbol for Gold. "It was a story about a vendor selling gold on the street corner and yelling, 'Hey You (Au) want to buy some Gold?' He pronounced Hey you as A u. After hearing that story, I've never forgotten that the symbol for Gold is Au. There's also an easy way to remember that the symbol for Silver is Ag. There are silver mines in **Ar**gentina. Ag is the symbol for Silver. Notice I made the letters Ag bold in **Arg**entina."

Guy said the hardest to learn will be Mercury, Hg. So, I'll just say it over and over again. Hg Mercury—Hg Mercury—Hg Mercury—Hg Mercury. The next hardest to learn will be Iron. So I'll say the symbol over and over again. Fe Iron—Fe Iron—Fe Iron. Then there's Copper, Cu. That's not hard, and Zn, Zinc is easy. I will learn them in no time."

Then Iron took out an index card with the symbols of the elements he thought important to remember. The elements Mercury, Zinc, Copper, Gold, Silver and Iron had created the card for Guy.

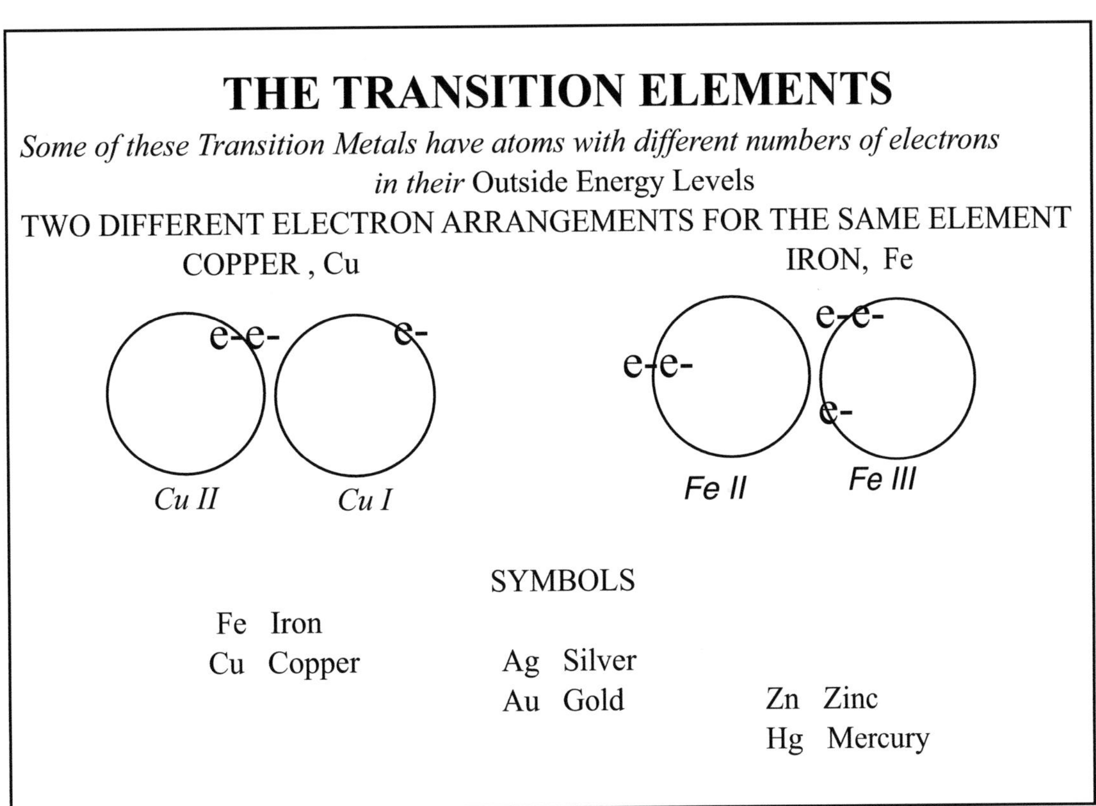

Guy was so thrilled and said, "This card will really help me, Thank you!"

With a little more knowledge of his world, Guy climbed into the bubble with Professor Terry. He pointed the magic wand in the direction of 3rd Street.and off they went. Just for the fun of it Guy made the bubble do some tricks. He went up high and then down low and made some sweeping circles. Guy wanted to have a little fun before continuing to meet the members of the Chemical Families. The Boron Family on 3rd Street was next.

THE BORON FAMILY
Chapter 5

Before long Guy spied the Boron Family flag and said, “Look Professor Terry their street is colored a light gray color. That must be their family color.” In a few minutes they arrived at 3rd Street. The home of the Boron Family. The sun was shining brightly as Professor Terry and Guy lowered the bubble onto the tarmac behind 3rd Street, Group 3A on the Periodic Table. This is where the Boron family lives, not far from the Transition Metals where they had just visited.

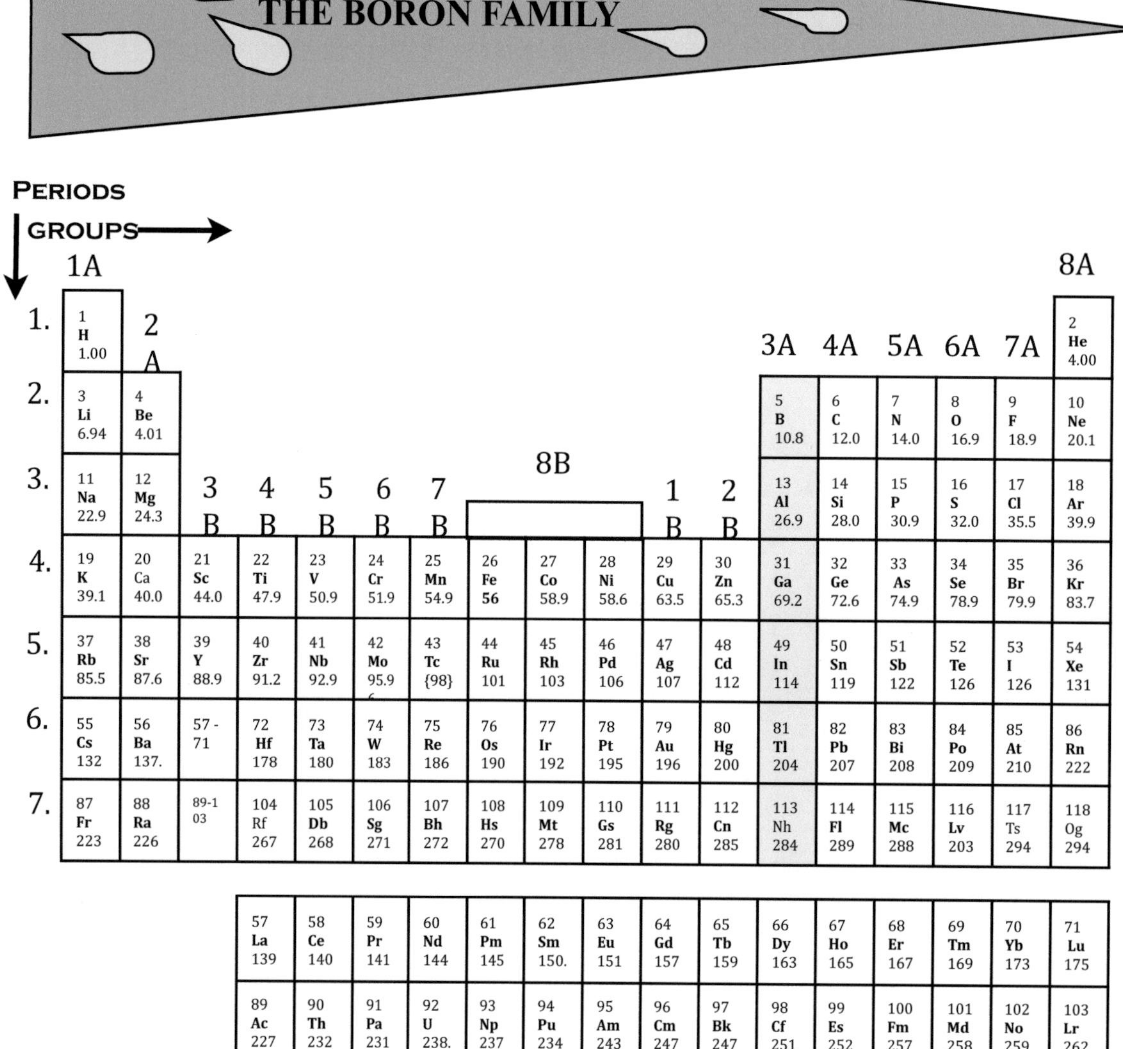

Periods	1A	2A	3B	4B	5B	6B	7B	8B	8B	8B	1B	2B	3A	4A	5A	6A	7A	8A
1.	1 H 1.00																	2 He 4.00
2.	3 Li 6.94	4 Be 4.01											5 B 10.8	6 C 12.0	7 N 14.0	8 O 16.9	9 F 18.9	10 Ne 20.1
3.	11 Na 22.9	12 Mg 24.3											13 Al 26.9	14 Si 28.0	15 P 30.9	16 S 32.0	17 Cl 35.5	18 Ar 39.9
4.	19 K 39.1	20 Ca 40.0	21 Sc 44.0	22 Ti 47.9	23 V 50.9	24 Cr 51.9	25 Mn 54.9	26 Fe 56	27 Co 58.9	28 Ni 58.6	29 Cu 63.5	30 Zn 65.3	31 Ga 69.2	32 Ge 72.6	33 As 74.9	34 Se 78.9	35 Br 79.9	36 Kr 83.7
5.	37 Rb 85.5	38 Sr 87.6	39 Y 88.9	40 Zr 91.2	41 Nb 92.9	42 Mo 95.9	43 Tc {98}	44 Ru 101	45 Rh 103	46 Pd 106	47 Ag 107	48 Cd 112	49 In 114	50 Sn 119	51 Sb 122	52 Te 126	53 I 126	54 Xe 131
6.	55 Cs 132	56 Ba 137.	57 - 71	72 Hf 178	73 Ta 180	74 W 183	75 Re 186	76 Os 190	77 Ir 192	78 Pt 195	79 Au 196	80 Hg 200	81 Tl 204	82 Pb 207	83 Bi 208	84 Po 209	85 At 210	86 Rn 222
7.	87 Fr 223	88 Ra 226	89-103	104 Rf 267	105 Db 268	106 Sg 271	107 Bh 272	108 Hs 270	109 Mt 278	110 Gs 281	111 Rg 280	112 Cn 285	113 Nh 284	114 Fl 289	115 Mc 288	116 Lv 203	117 Ts 294	118 Og 294

57 La 139	58 Ce 140	59 Pr 141	60 Nd 144	61 Pm 145	62 Sm 150.	63 Eu 151	64 Gd 157	65 Tb 159	66 Dy 163	67 Ho 165	68 Er 167	69 Tm 169	70 Yb 173	71 Lu 175
89 Ac 227	90 Th 232	91 Pa 231	92 U 238.	93 Np 237	94 Pu 234	95 Am 243	96 Cm 247	97 Bk 247	98 Cf 251	99 Es 252	100 Fm 257	101 Md 258	102 No 259	103 Lr 262

3rd Street in Periodic Table Land
Guy Meets the Boron Family

Boron met them by the flagpole where he was busy adjusting the ropes that tied the family flag in place. Boron greeted them and said, “I hope you noticed our flag and street are both painted a silvery gray color. That’s our family color.” Boron confirmed what Guy had guessed to be true.

Boron continued, “Gray is our family color because we are all silvery metals. We drew pictures of whales on our silvery flag because they are gray like us. We love our family color as much as we love those magnificent whales. The word spread that you were coming. So we’re ready to tell you more about our family. Here’s our name patch.”

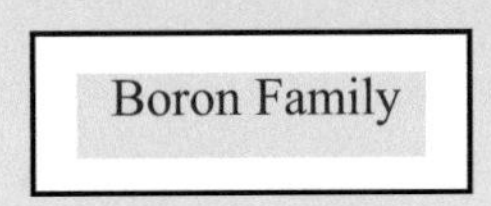

“Let me tell you more about our family. I am the only different kind of element in my family. I’m a metalloid. I can act like a metal or a non-metal. The rest of the family members are metals. Since we’re located right after the Transition metals on the Periodic Table, our metals are called Post Transition Metals. Post means after. That was a good name for our metals because indeed we are right after the Transition Metals on the Periodic Table.”

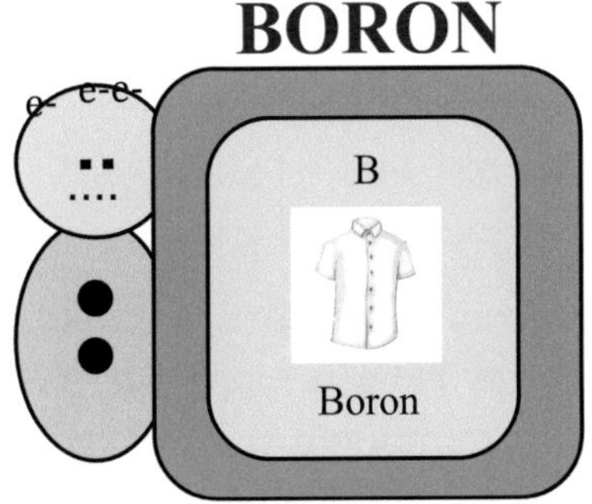

“I’m **BORON**. I am used to strengthen glass to keep it from breaking easily. They add me to fiberglass to make it insulate your house better. One of my products, Borax is a good cleaner when added to laundry detergent, clothes come out whiter. I will draw a white shirt because I get white clothes clean and bright.”

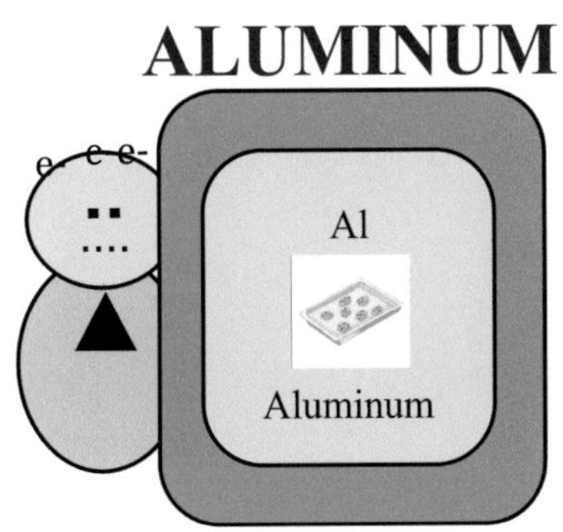

“I’m **ALUMINUM**. I’m a light metal. Almost everyone knows me. I’m used to make many things from airplanes to kitchen utensils—anything that requires a light metal. Cookie sheets are my favorite use. I love chocolate chip cookies that people bake on me. This is what I’ll draw on my patch.”

GALLIUM

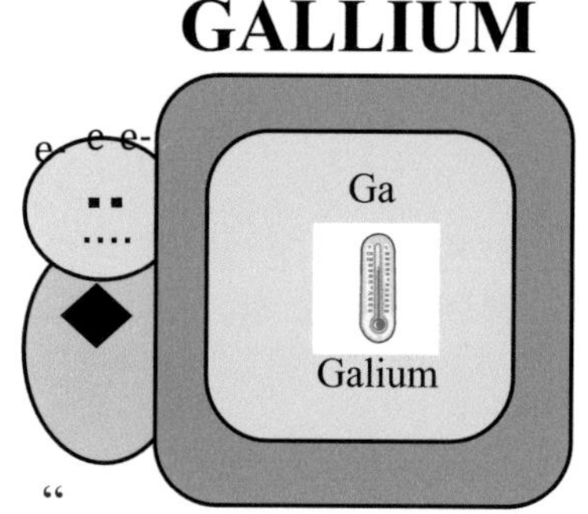

GALLIUM said, “My uses are associated with my low melting point. I can melt in your hand at normal body temperature which is about 98^0 F. Compare this to Aluminum which melts at over 600^0 F. I'm also used in semiconductor applications. The drawing on my patch will show my best use in hight temperature thermometers because my liquid can also withstand high temperatures.”

INDIUM

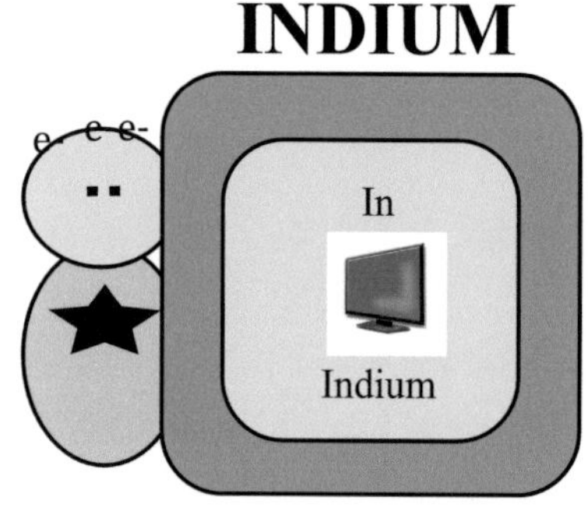

“I'm **INDIUM**. I'm used in soldering, cryogenics, to make semiconductors and solar panels. Solar panels use the sun to make electricity. My greatest demand today is for my compound Indium Tin Oxide used in LCD (Liquid Crystal Displays), for touch screen and flat screen TVs and also for solar panels. For my element patch I'm going to draw a TV.

THALLIUM

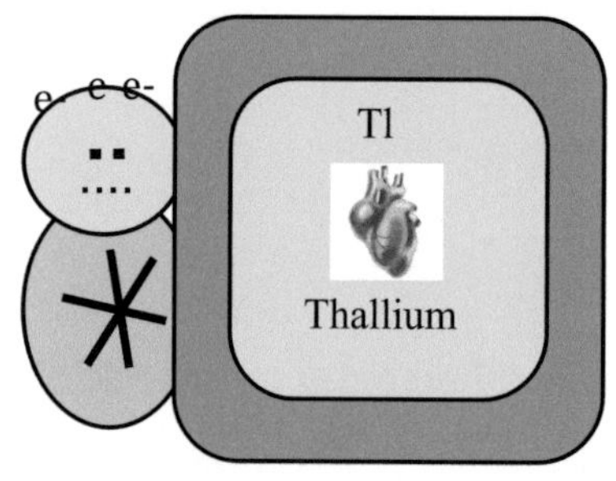

THALLIUM said, “My best use is in medicine. I'm used by cardiologists in performing stress tests to evaluate the functioning of the human heart. I will draw a heart to show my best use.”

Aluminum said, “You met all the Boron Family elements except one. Did you notice that element Nihonium-Nh, Atomic Number 113 did not have a patch for our banner? It's because he's very radioactive and a newly discovered element. Also he is made in the lab and stays in existence for only seconds. No surprise, he doesn't have any uses.”

Just then, all five elements in the Boron Family came running to give an index card to Guy with the information he had learned about the Boron Family.They all liked Guy and wanted him to remember them.

Guy here's a card we made up for you so that you can remember us. It has lots of information on it about the elements in our family. We would like you to notice most of all that our family members all have 3 electrons in our Outside Energy Level.

CHEMICAL FAMILIES
GROUP 3A
THE BORON FAMILY

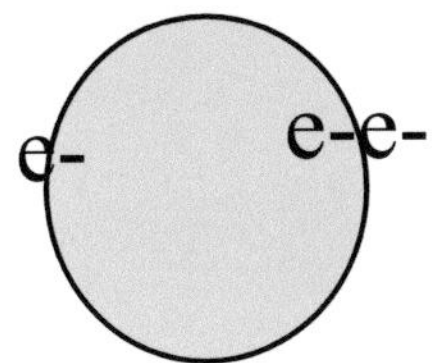

Group 3A each element has 3 electrons in its Outside Energy Level.

Boron	(B)	used in Borax, a cleaning product.
Aluminum	(Al)	used in airplanes and cookie sheets.
Gallium	(Ga)	used in high temperature thermometers.
Indium	(In)	used in LCD TVs and solar panels.
Thallium	(Tl)	used by heart doctors in stress tests.

Guy looked it over, thanked the little elements and told them he would put their very special card with all his important papers where it would be safe. "I'll learn what you have on this card and remember all about you until we meet again."

Professor Terry and Guy hopped into their wonderful means of transportation and were on their way to visit the Carbon Family. On the way they did not miss a single part of the beautiful landscape below. There was a large willow tree with its branches reaching so low the leaves dipped into the cool lake water. In the grassy area away from the lake were some adorable fluffy white baby lambs grazing. In the distance Guy saw a little element out for a horseback ride. It looked like he was headed home.

Professor Terry brought Guy's attention back to his job at hand, pointing the magic bubble in the right direction. "We're almost there now Guy get that magic wand working. Steer us in the right direction. We're almost close enough to see the Carbon family's flag now.

Guy felt like zooming down to the grassy field and playing with those adorable white fluffy lambs. However, Professor Terry kept him on tract. They still had five Chemical Families to visit.

THE CARBON FAMILY
Chapter 6

Guy said, "From up here it looks like their family color is black." They had to be a little closer to be sure. As they got closer to 4th Street, they could see the little black elements moving from house to house, busy as can be. It was obvious they were coordinating plans for the festival. Now they were close enough to not only see that Carbon's flag was black but they could read the words on the flag, The Carbon Family.

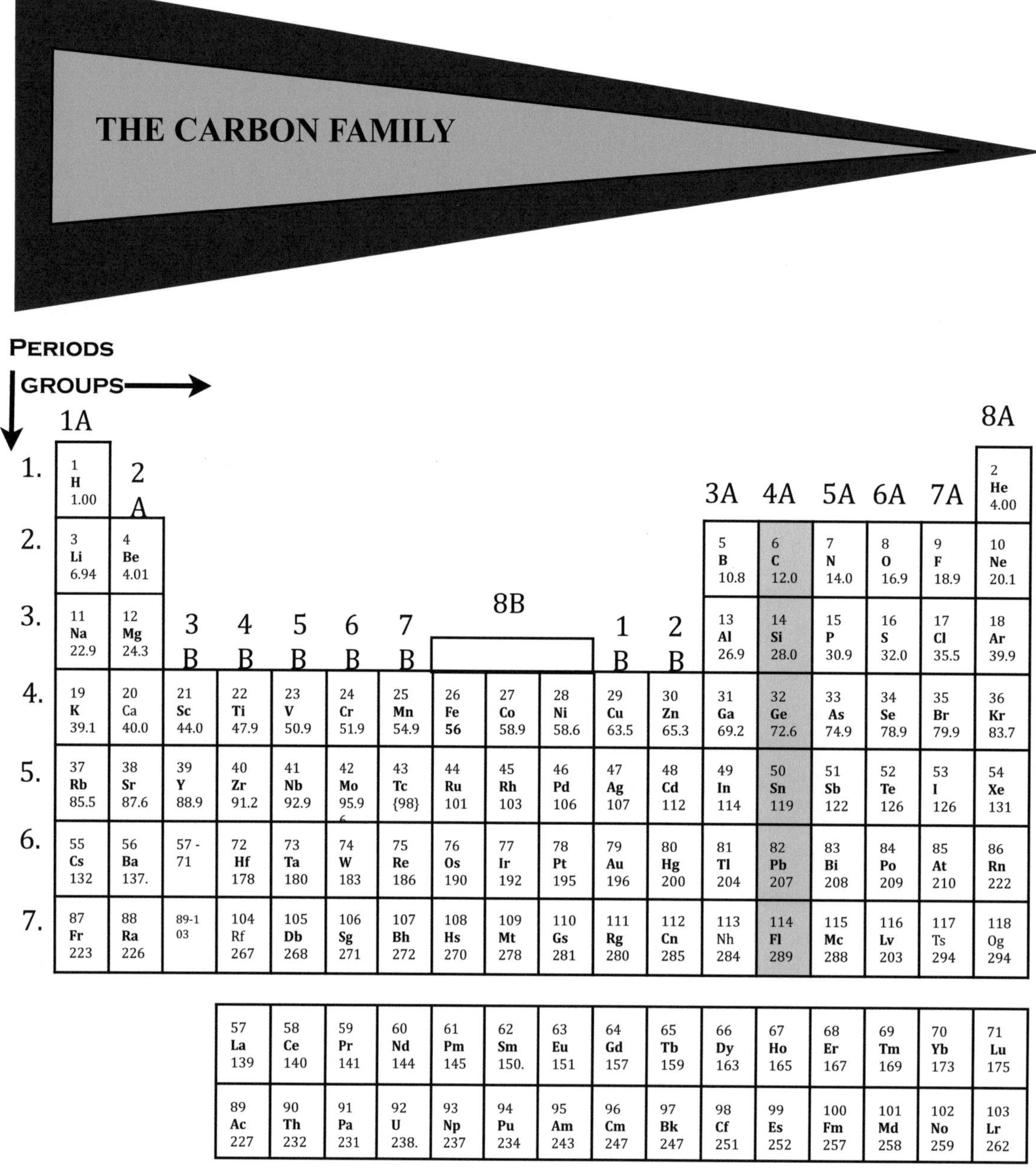

	1A	2A	3B	4B	5B	6B	7B	8B	8B	8B	1B	2B	3A	4A	5A	6A	7A	8A
1.	1 H 1.00																	2 He 4.00
2.	3 Li 6.94	4 Be 4.01											5 B 10.8	6 C 12.0	7 N 14.0	8 O 16.9	9 F 18.9	10 Ne 20.1
3.	11 Na 22.9	12 Mg 24.3											13 Al 26.9	14 Si 28.0	15 P 30.9	16 S 32.0	17 Cl 35.5	18 Ar 39.9
4.	19 K 39.1	20 Ca 40.0	21 Sc 44.0	22 Ti 47.9	23 V 50.9	24 Cr 51.9	25 Mn 54.9	26 Fe 56	27 Co 58.9	28 Ni 58.6	29 Cu 63.5	30 Zn 65.3	31 Ga 69.2	32 Ge 72.6	33 As 74.9	34 Se 78.9	35 Br 79.9	36 Kr 83.7
5.	37 Rb 85.5	38 Sr 87.6	39 Y 88.9	40 Zr 91.2	41 Nb 92.9	42 Mo 95.9	43 Tc {98}	44 Ru 101	45 Rh 103	46 Pd 106	47 Ag 107	48 Cd 112	49 In 114	50 Sn 119	51 Sb 122	52 Te 126	53 I 126	54 Xe 131
6.	55 Cs 132	56 Ba 137.	57 - 71	72 Hf 178	73 Ta 180	74 W 183	75 Re 186	76 Os 190	77 Ir 192	78 Pt 195	79 Au 196	80 Hg 200	81 Tl 204	82 Pb 207	83 Bi 208	84 Po 209	85 At 210	86 Rn 222
7.	87 Fr 223	88 Ra 226	89-103	104 Rf 267	105 Db 268	106 Sg 271	107 Bh 272	108 Hs 270	109 Mt 278	110 Gs 281	111 Rg 280	112 Cn 285	113 Nh 284	114 Fl 289	115 Mc 288	116 Lv 203	117 Ts 294	118 Og 294

57 La 139	58 Ce 140	59 Pr 141	60 Nd 144	61 Pm 145	62 Sm 150.	63 Eu 151	64 Gd 157	65 Tb 159	66 Dy 163	67 Ho 165	68 Er 167	69 Tm 169	70 Yb 173	71 Lu 175
89 Ac 227	90 Th 232	91 Pa 231	92 U 238.	93 Np 237	94 Pu 234	95 Am 243	96 Cm 247	97 Bk 247	98 Cf 251	99 Es 252	100 Fm 257	101 Md 258	102 No 259	103 Lr 262

Guy Meets the Carbon Family 4th Street in Periodic Table Land

Carbon met them at the flagpole. “So glad to see you again,” said Carbon. “It’s been a while since you drew my Bohr model. I guess you have heard about the contest. Our family color is black but on those surfaces that we have to put writing we have to use a gray color so you can see the writing. You will enjoy meeting the members of my family. They will be eager to show you the patch they’re preparing for the contest. I made this patch for the top of the banner showing our family name. Here it is.”

CARBON began to talk about himself. “People study me for a full year just to learn the basics of how I form compounds. It’s called Organic Chemistry. Carbon is in plastics, butter gasoline, wood, natural gas, and many more products. As a normal part of breathing, people exhale my compound carbon dioxide. Leaves on trees take in the carbon dioxide and use it for Photosynthesis. Carbon is found in all living things. It’s also found in its natural form as coal which is used for fuel and as a diamond, known as a girl’s best friend. I will draw a diamond for my patch.”

SILICON said, “I’m used for making glass. I’m used in computers and inside television sets. I’m part of sand that is the main ground cover in deserts. I line the edges of the oceans with beaches that are vacation spots. Kids love to go to these beaches and build sand castles. Sand is really my compound, silicon dioxide. I’m drawing a sand castle on my patch.”

GERMANIUM

GERMANIUM said, "I'm a metalloid. I am mainly used as a semiconductor in electronics and in fiber optics. I am also used in wide angle camera lenses and in microscope lenses. I'll be drawing a camera on my patch."

TIN

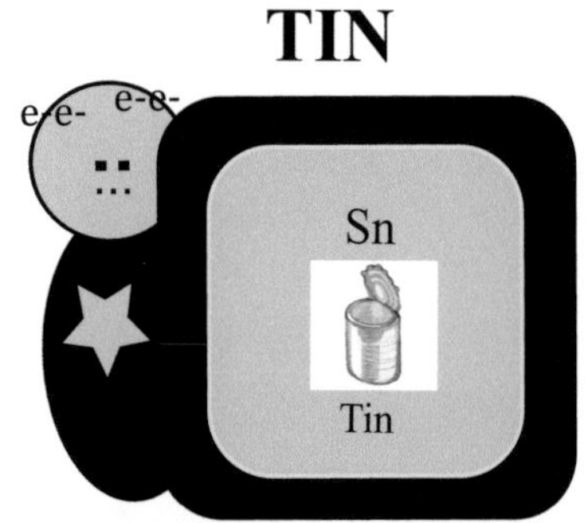

"I'm **TIN**," said the element that Guy visited next. "I am a real metal. I'm soft silvery white. I've been around for 5,500 years. To this day, people call the cans that food comes in *tin cans* even if they are made of other metals. Since I resist corrosion, they use me to coat other metals to prevent rusting. I am going to draw a tin can on my square, and my tin can is really made of the element tin."

LEAD

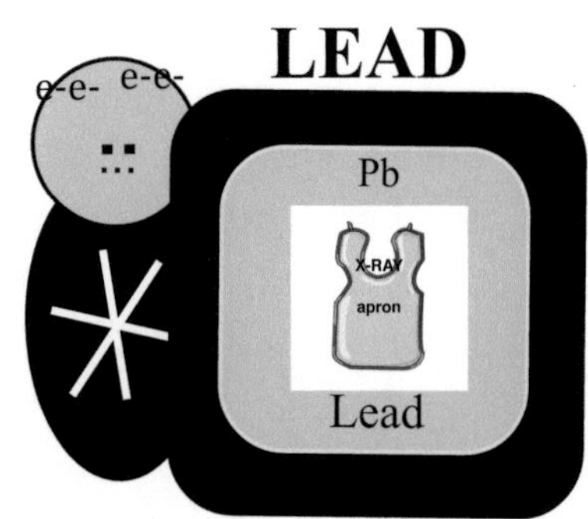

"I am **Lead**—a real metal. For years I was used in plumbing. Notice that my symbol is Pb letters in the word **p**lum**b**ing. Now that people have learned that I am poison to the human body, they have stopped using me in pipes and also paints which was another use for me that I enjoyed. One use I have today is as a shield against radiation.Your dentist uses a lead apron as a protection against radiation when he takes X-rays of your teeth.

Carbon said, "Now you've met the members of my family. However, there's one more element in Group 4A. It's the element, Flerovium-Fl, Atomic Number 114." So Carbon said, "Let me tell you about Flerovium. A scientist created just one atom of Flerovium in his laboratory. Flerovium is a superheavy artificial radioactive element and so far no uses for that element have been found."

Professor Terry told Guy, "All the elements from 113 to 118 are radioactive. They have been created in the laboratory and stay in existence for only seconds or less. But it is good to know that these elements do exist."

Then Carbon said, "Before you go, I have to tell you something that is different about me. Carbon atoms have a property that is different from all the other elements. My atoms can link together with other Carbon atoms sometimes in long chains. When you write my symbol you use sticks instead of e^- for the electrons Here's how Carbon can look when the Carbon atoms link together.

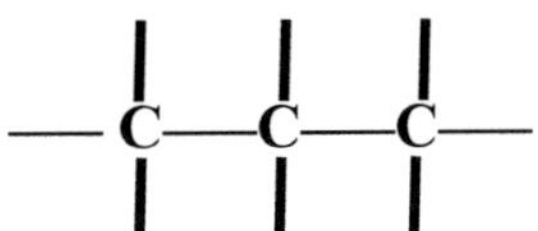

These sticks represent Carbon's 4 electrons. This is where other Carbon atoms and other elements join together to form compounds. Carbon and these elements share electrons. When Hydrogen joins on to these electron sticks, the compounds are called Hydrocarbons. Learning about these compounds is fascinating. There is a whole branch of chemistry dedicated to the study of Carbon and its compounds. It is called Organic Chemistry."

At that moment members of the Carbon family came running over to Guy with a large index card just like ones the elements in other families had made for Guy. By this time Guy was not surprised to get a card. However, he was still delighted that these little elements took the time to make a card for him.

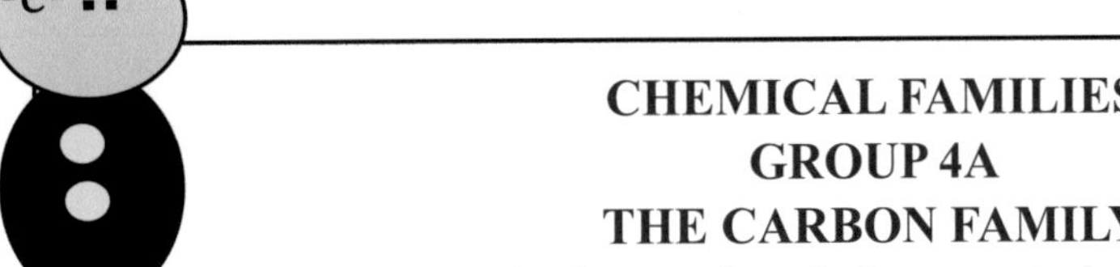

CHEMICAL FAMILIES
GROUP 4A
THE CARBON FAMILY

In Group 4A, each element has 4 electrons in its Outside Energy Level

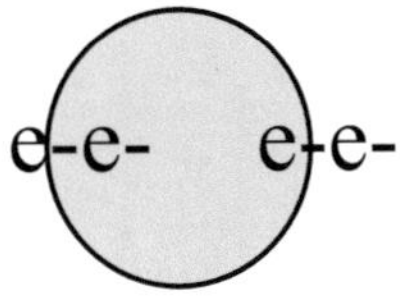

6	Carbon	(C)	found in all living things.
14	Silicon	(Si)	found in sand.
32	Germanium	(Ge)	made transistors possible.
50	Tin	(Sb)	makes bronze with copper and zinc.
82	Lead	(Pb)	Lead aprons shield us from radiation.

Then they were off to 5th Street the home of the Nitrogen Family.

THE NITROGEN FAMILY
Chapter 7

When Professor Terry and Guy arrived at 5th Street, Guy admired the family flag hanging high over the entrance. He liked especially the flag's bright red color. "One look at the pentagon with its 5 sides," said Guy, "and anyone would know that this was the 5th Street flag. The N inside the pentagon gives a clue that the 5th Street flag belongs to the Nitrogen Family. How clever!

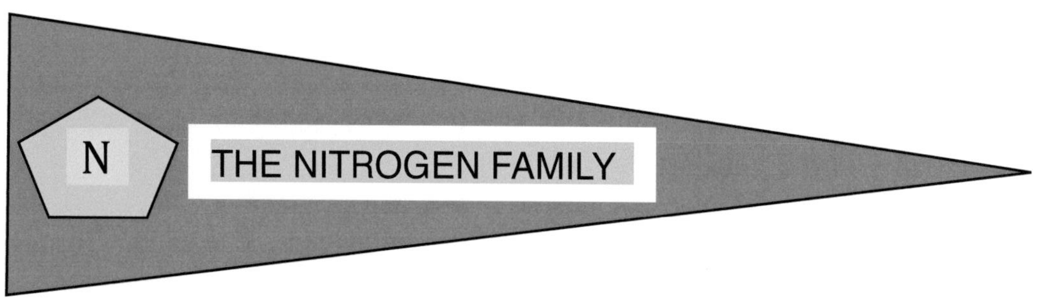

PERIODS

GROUPS→

THE PERIODIC TABLE OF THE ELEMENTS

PERIODS	1A	2A	3B	4B	5B	6B	7B	8B	8B	8B	1B	2B	3A	4A	5A	6A	7A	8A
1.	1 H 1.00																	2 He 4.00
2.	3 Li 6.94	4 Be 4.01											5 B 10.8	6 C 12.0	7 N 14.0	8 O 16.9	9 F 18.9	10 Ne 20.1
3.	11 Na 22.9	12 Mg 24.3											13 Al 26.9	14 Si 28.0	15 P 30.9	16 S 32.0	17 Cl 35.5	18 Ar 39.9
4.	19 K 39.1	20 Ca 40.0	21 Sc 44.0	22 Ti 47.9	23 V 50.9	24 Cr 51.9	25 Mn 54.9	26 Fe 56	27 Co 58.9	28 Ni 58.6	29 Cu 63.5	30 Zn 65.3	31 Ga 69.2	32 Ge 72.6	33 As 74.9	34 Se 78.9	35 Br 79.9	36 Kr 83.7
5.	37 Rb 85.5	38 Sr 87.6	39 Y 88.9	40 Zr 91.2	41 Nb 92.9	42 Mo 95.9 6	43 Tc {98}	44 Ru 101	45 Rh 103	46 Pd 106	47 Ag 107	48 Cd 112	49 In 114	50 Sn 119	51 Sb 122	52 Te 126	53 I 126	54 Xe 131
6.	55 Cs 132	56 Ba 137.	57 - 71	72 Hf 178	73 Ta 180	74 W 183	75 Re 186	76 Os 190	77 Ir 192	78 Pt 195	79 Au 196	80 Hg 200	81 Tl 204	82 Pb 207	83 Bi 208	84 Po 209	85 At 210	86 Rn 222
7.	87 Fr 223	88 Ra 226	89-1 03	104 Rf 267	105 Db 268	106 Sg 271	107 Bh 272	108 Hs 270	109 Mt 278	110 Gs 281	111 Rg 280	112 Cn 285	113 Nh 284	114 Fl 289	115 Mc 288	116 Lv 203	117 Ts 294	118 Og 294

57 La 139	58 Ce 140	59 Pr 141	60 Nd 144	61 Pm 145	62 Sm 150.	63 Eu 151	64 Gd 157	65 Tb 159	66 Dy 163	67 Ho 165	68 Er 167	69 Tm 169	70 Yb 173	71 Lu 175
89 Ac 227	90 Th 232	91 Pa 231	92 U 238.	93 Np 237	94 Pu 234	95 Am 243	96 Cm 247	97 Bk 247	98 Cf 251	99 Es 252	100 Fm 257	101 Md 258	102 No 259	103 Lr 262

Guy Meets the Nitrogen Family
5th Street in Periodic Table Land

While knocking at Nitrogen's door, they admired the shiny brass plaque displaying Nitrogen's Atomic Number 7. As they were waiting, Professor Terry explained to Guy that 5th Street is the beginning of the elements that are non-metals. "On 5th Street the most important non-metals for you to remember are Nitrogen and Phosphorus. You have already met them when you drew their Bohr models."

At that moment the door swung open, and Nitrogen reached out and shook their hands with a warm welcome. Nitrogen said, "I bet you are the judges for the Banner Contest!"

"Not really," said Professor Terry, "But we do think that any judge would like your flag. It is so symbolic. Good job! We've come to meet the members of your family."

Nitrogen invited them in, and showed them the family name patch that was going to be put on the top of their family banner.

Nitrogen told them all elements were doing their best to make winning patches for their banner Then remembering why they had come said," Let's begin learning about my family right now. Come with me."

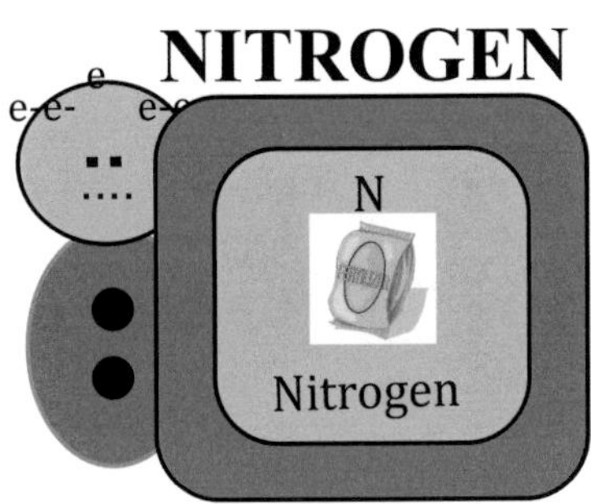

NITROGEN began talking about himself. "So many people think of air as being pure Oxygen. However, only 20% of the air is Oxygen. Nitrogen makes up 78% of the earth's atmosphere. That's me! I am colorless, odorless and tasteless, and I'm only slightly reactive. My compounds are called nitrides, nitrates and nitrites. Some of my compounds are used as fertilizers. Some plants like clover put nitrogen back into the soil to make farm land more fertile. I'm drawing a bag of fertilizer on my square."

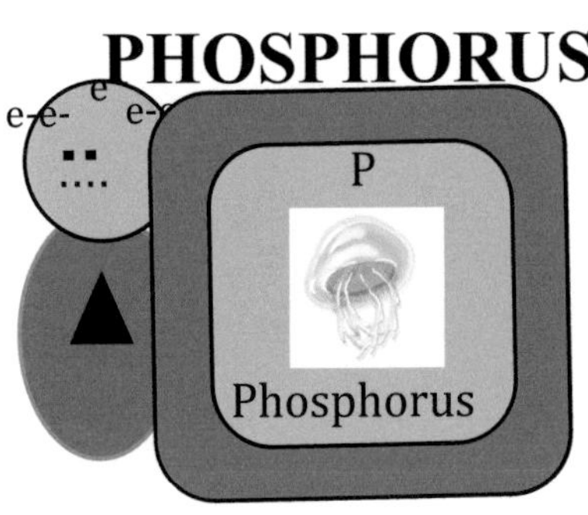

PHOSPHORUS is essential for life. It's part of our DNA, and its compounds are used as fertilizers. "I am an interesting element. I come in two forms: red and white Phosphorus. White Phosphorus is better known than red Phosphorus which is found mainly on match boxes. White Phosphorus possesses the quality of luminescence. In the presence of Oxygen, white Phosphorus can glow. Some sea creatures are phosphorescent. That means they glow. So I'm drawing a jelly fish in my square."

Preparing Guy to meet Arsenic and Antimony who were next on the list to visit, Nitrogen said, "Both Arsenic and Antimony are metalloids. They can either act like a metal or a non-metal in forming compounds. We'll move on and let them tell you their story."

ARSENIC

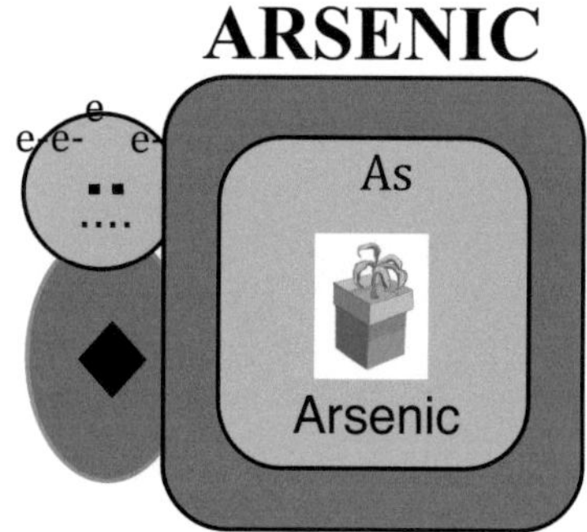

ARSENIC said,"I have lots of uses because I'm a poison. I'm used as a herbicide That means I kill weeds. I'm also an insecticide. I get rid of insects. I'm also used to control the rodent population. Someone even wrote a play about me called *Arsenic and Old Lace.* I'm drawing a wilting weed in my square."

ANTIMONY

ANTIMONY said, "I've been around since ancient times. Since way back, people have used me in cosmetics. During colonial days, dishes were made of pewter, one of my alloys. I'm used to vulcanize rubber that's used for tires. My latest use is in microelectronics. For my square I will be drawing a modern day lady using cosmetics."

BISMUTH

BISMUTH is a post-transition metal. it's a lot like Arsenic and Antimony in its chemical structures and properties. Bismuth said, "When you buy lipsticks with a frosty look, you can thank me. My best use is in helping people deal with the discomforts of the digestive system. The most common use for Bismuth is as an antacid," Bismuth said, "I'm going to draw a picture of a bottle of antacid for my square."

Guy said, "I noticed that you did not tell me about element 115." Nitrogen explained, " Element 115 was just recently named. It's called Moscovium-Mc. It was named after the location of the lab in which it was discovered. That was Moscow, Russia. Moscovium is a radioactive element, and so far it has no known uses. It is an unstable element and only stays in existence less than a second."

Nitrogen said, "If you want to remember one important fact about the Nitrogen Family, remember we have 5 electrons in our Outside Energy Level. It will be important when you learn about how we form compounds.

Just then, all the Nitrogen elements came rushing down to see Guy saying, "Hey. Guy, we don't ever want you to forget us. So here's a card we made up for you."

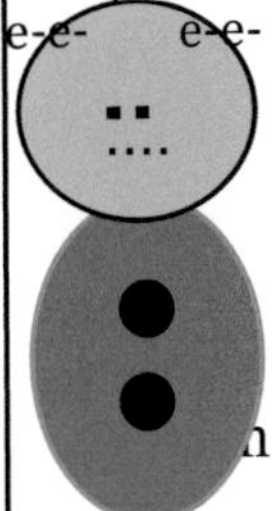

CHEMICAL FAMILIES
GROUP 5A
THE NITROGEN FAMILY
n Group 5A each element has 5 electrons in its Outside Energy Level

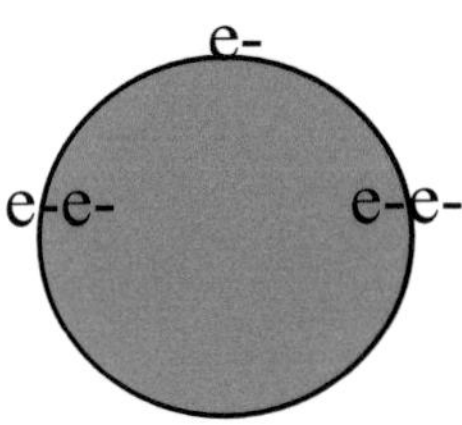

The first two elements behave like non-metals.
7 Nitrogen (N) A gas making up about 78 % of our air.
15 Phosphorous (P) Used in matches and fertilizers.
The last three are different; they behave like metals.
33 Arsenic (As) is a rat poison.
51 Antimony (Sb) is used in cosmetics.
83 Bismuth (Bi) is used as an antacid.

Guy answered, "Don't worry Nitrogen Family I'll remember you. Professor Terry and Guy waved good bye to the Nitrogens and were on their way. Guy turned to the professor and said, "I'll especially remember Phosphorus. He was really different coming in two forms. I liked how he makes sea creatures glow in the dark."

THE OXYGEN FAMILY
Chapter 8

The sun was high in the sky and shining brightly. Guy took one last look at the Nitrogen family below. All those beautiful flowers looked so pretty. Now he knew it was because Nitrogen and Phosphorus knew how to fertilize their gardens.

They soon focused on the next family they were headed to see on 6th Street. Professor Terry, looking into the distance remarked, "We're almost there. The clouds parted, and Guy said, "I can see the flag now. It is a pretty yellow color with stars on it, and the street is yellow, too"

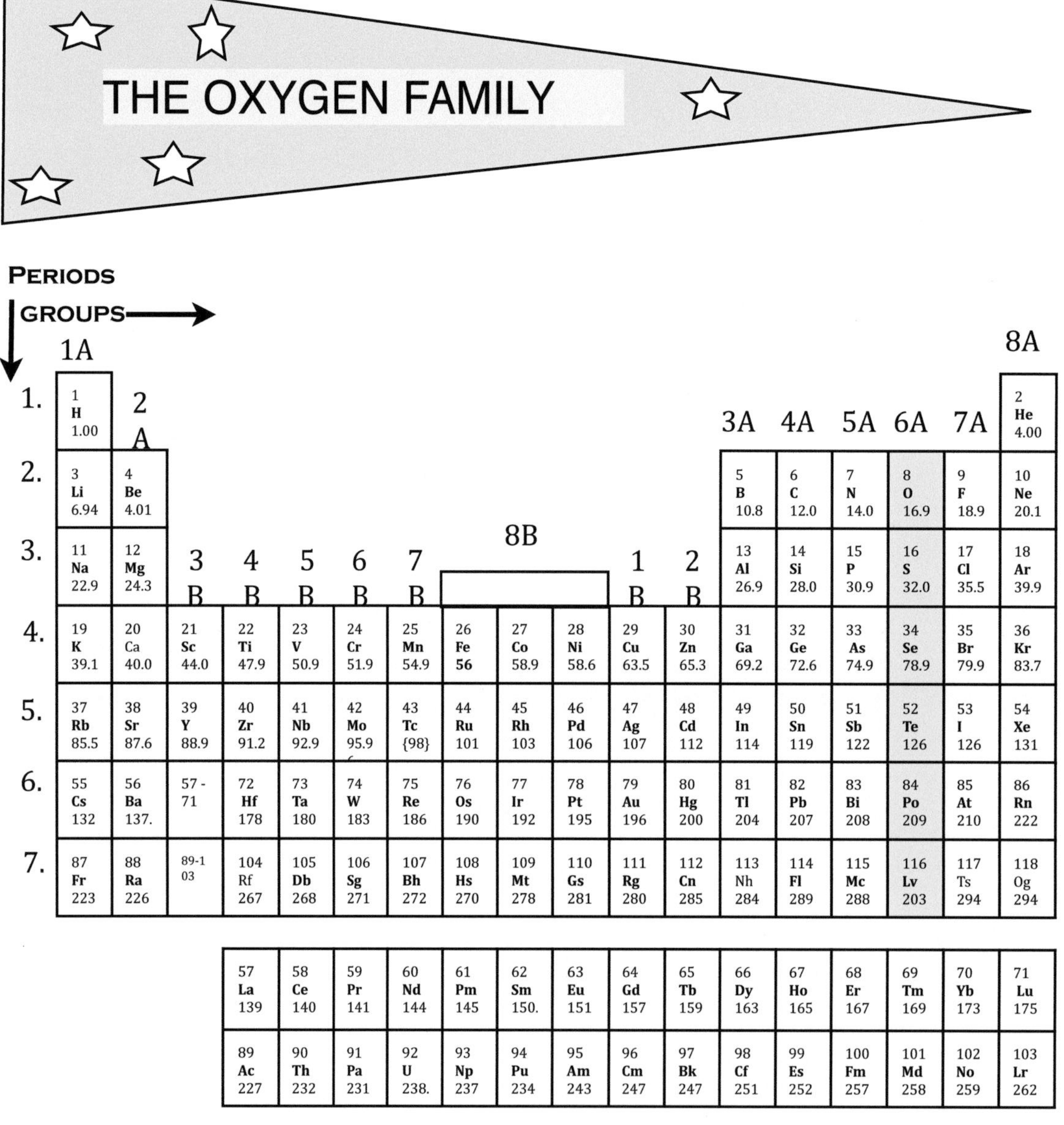

Periods	1A	2A	3B	4B	5B	6B	7B	8B			1B	2B	3A	4A	5A	6A	7A	8A
1.	1 **H** 1.00																	2 **He** 4.00
2.	3 **Li** 6.94	4 **Be** 4.01											5 **B** 10.8	6 **C** 12.0	7 **N** 14.0	8 **O** 16.9	9 **F** 18.9	10 **Ne** 20.1
3.	11 **Na** 22.9	12 **Mg** 24.3											13 **Al** 26.9	14 **Si** 28.0	15 **P** 30.9	16 **S** 32.0	17 **Cl** 35.5	18 **Ar** 39.9
4.	19 **K** 39.1	20 Ca 40.0	21 **Sc** 44.0	22 **Ti** 47.9	23 **V** 50.9	24 **Cr** 51.9	25 **Mn** 54.9	26 **Fe** **56**	27 **Co** 58.9	28 **Ni** 58.6	29 **Cu** 63.5	30 **Zn** 65.3	31 **Ga** 69.2	32 **Ge** 72.6	33 **As** 74.9	34 **Se** 78.9	35 **Br** 79.9	36 **Kr** 83.7
5.	37 **Rb** 85.5	38 **Sr** 87.6	39 **Y** 88.9	40 **Zr** 91.2	41 **Nb** 92.9	42 **Mo** 95.9	43 **Tc** {98}	44 **Ru** 101	45 **Rh** 103	46 **Pd** 106	47 **Ag** 107	48 **Cd** 112	49 **In** 114	50 **Sn** 119	51 **Sb** 122	52 **Te** 126	53 **I** 126	54 **Xe** 131
6.	55 **Cs** 132	56 **Ba** 137.	57 - 71	72 **Hf** 178	73 **Ta** 180	74 **W** 183	75 **Re** 186	76 **Os** 190	77 **Ir** 192	78 **Pt** 195	79 **Au** 196	80 **Hg** 200	81 **Tl** 204	82 **Pb** 207	83 **Bi** 208	84 **Po** 209	85 **At** 210	86 **Rn** 222
7.	87 **Fr** 223	88 **Ra** 226	89-1 03	104 Rf 267	105 **Db** 268	106 **Sg** 271	107 **Bh** 272	108 **Hs** 270	109 **Mt** 278	110 **Gs** 281	111 **Rg** 280	112 **Cn** 285	113 Nh 284	114 **Fl** 289	115 **Mc** 288	116 **Lv** 203	117 Ts 294	118 Og 294

57 **La** 139	58 **Ce** 140	59 **Pr** 141	60 **Nd** 144	61 **Pm** 145	62 **Sm** 150.	63 **Eu** 151	64 **Gd** 157	65 **Tb** 159	66 **Dy** 163	67 **Ho** 165	68 **Er** 167	69 **Tm** 169	70 **Yb** 173	71 **Lu** 175
89 **Ac** 227	90 **Th** 232	91 **Pa** 231	92 **U** 238.	93 **Np** 237	94 **Pu** 234	95 **Am** 243	96 **Cm** 247	97 **Bk** 247	98 **Cf** 251	99 **Es** 252	100 **Fm** 257	101 **Md** 258	102 **No** 259	103 **Lr** 262

Guy Meets the Oxygen Family
6th Street in Periodic Table Land

Still floating above 6th Street in the magic bubble Professor Terry told Guy, "The family on 6th Street is called the Oxygen Family, and Oxygen is the first element on 6th Street."

Guy said, "That makes sense. Since Boron on 3rd Street, all the families were named after the first element in the family: Boron, The Boron Family; Carbon, The Carbon Family; Nitrogen, The Nitrogen Family. That's the way it's been going."

When they landed, Oxygen was right there to greet them. Oxygen explained about the family color. "We chose yellow as our family color to honor Sulfur, a member of our family. He is a yellow solid, sometimes a yellow powder." I'm so glad you stopped by to see my family. Here's the family rectangle I made. We will put it at the top of our banner for the contest.

The Oxygen Family

Guy said, "I noticed you painted 6th Street yellow. Your family's special color. looks so good. I'd like to meet all the members of your family."

Oxygen said, 'That's great. All the elements will be eager to show you the patch they are making for the contest. Sit down, and I'll tell you all about my element and my patch. I'm quite a busy element with all the festivities. Let me begin."

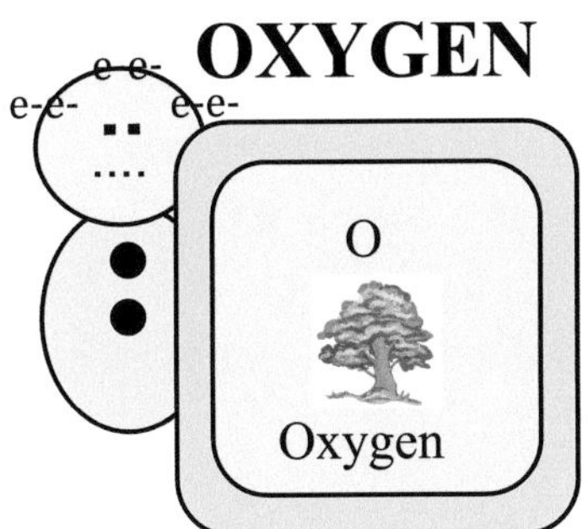

" Now I'd like to show you the square I'm making for our banner," said **Oxygen.** "I drew a tree on it because trees using sunlight create more Oxygen for our atmosphere. People need Oxygen to live. I'm essential to life on earth. People with respiratory disease use medical Oxygen in tanks. In airplanes Oxygen is available if the passenger cabin loses pressure. Also, you need to know something different about me, I'm diatomic. That means I like to travel with another Oxygen atom. To show this we write Oxygen as O_2"

Then, Oxygen led Guy down 6th Street to meet Sulfur. They looked at all the pretty flowers that were growing along. the path down 6th Street. It reminded Guy of how much he loved nature–flowers, trees, rocks as well as his favorite, the night sky.

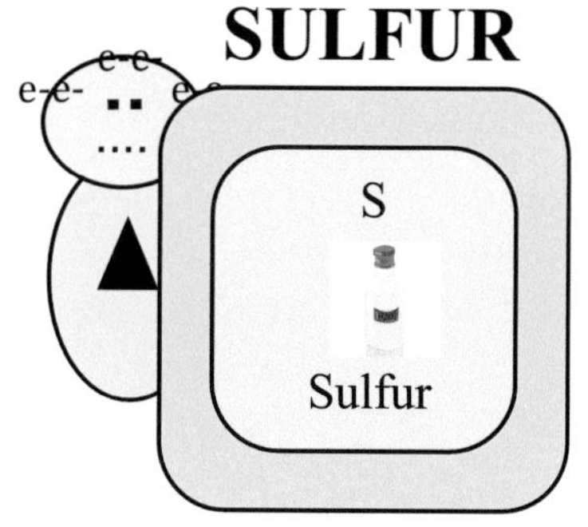

SULFUR began, "I'm a yellow solid, a non-metal. Some of my compounds are called sulfides, others sulfates, or sulfites. Sulfur was the best drug to cure infections before antibiotics were discovered. In many parts of our country, water has a terrible odor when sulfur is present in the ground in the form of hydrogen sulfide. Some people call it egg water because it smells like rotten eggs. I can be made into a strong acid, known as sulfuric acid. That's what I'm drawing—a bottle of sulfuric acid."

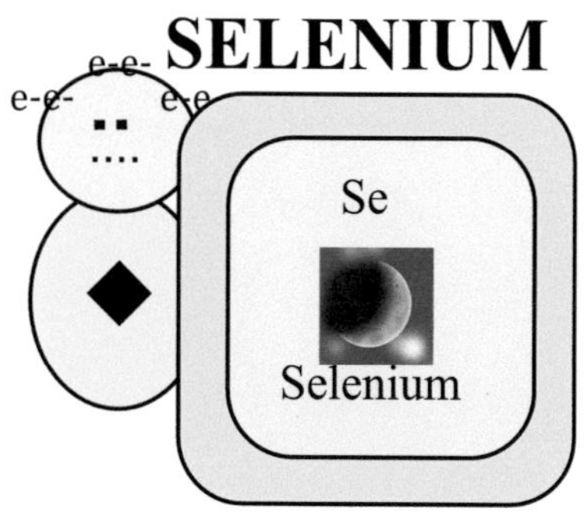

SELENIUM shared that he was a non-metal and had many uses. "My main commercial use today is in glassmaking and in pigments. I'm a semiconductor and used in photocells. People use me in DC power surge protectors to protect their computers. Also, my name is from a Greek word for the moon. I'm drawing a moon for my patch."

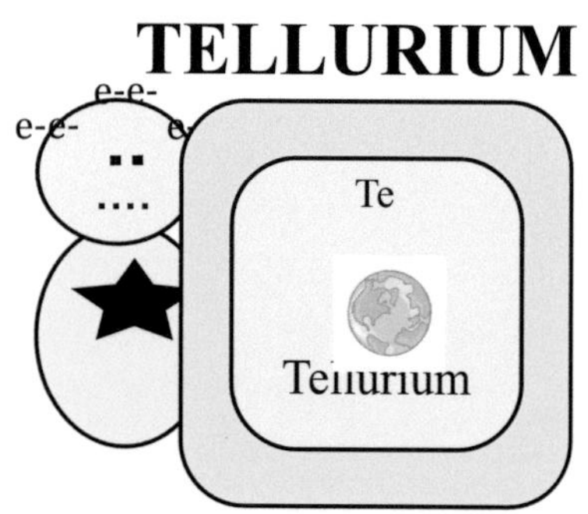

TELLURIUM was happy to see Guy, and Guy was happy to finally meet Tellurium. This was the element whose name he had identified on a quiz. Tellurium shared, "I was discovered in 1783, but no one has found many uses for me yet. I am in short supply on earth. Because tellurium gives a garlic odor to breath, it has helped detectives identify the cause of death as tellurium poisoning." Tellurium continued, "I'm a metal. I am a silvery white color. I was named for the Latin word *tellus* that means Earth. I'm drawing the Earth on my patch."

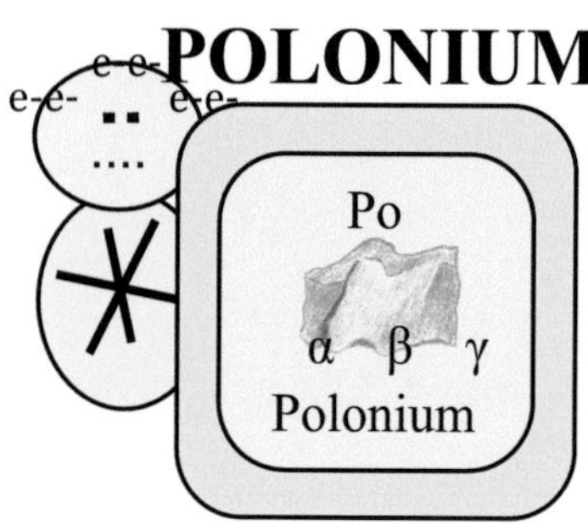

POLONIUM: Pierre and Marie Curie discovered Polonium as they were studying Uranium ore. Working with the alpha particles, given off by Polonium cost them their lives. Before that, it was not known that alpha particles were poisonous. Since then scientists wear protective clothing while working with radioactive materials. "I'm drawing a radioactive rock for my square showing alpha, beta and gamma radiation."

Oxygen said, “In Group 6A we have one more element Livermorium-Lv, Atomic Number 116 It is extremely radioactive, and it is not found in nature. It has only been developed in the laboratory by scientists and has no uses.”

“Now I’ve told you about all the elements in my family.”

As Oxygen finished talking, all of the Oxygen family elements ran up to Guy holding the Oxygen family card that they made for him.

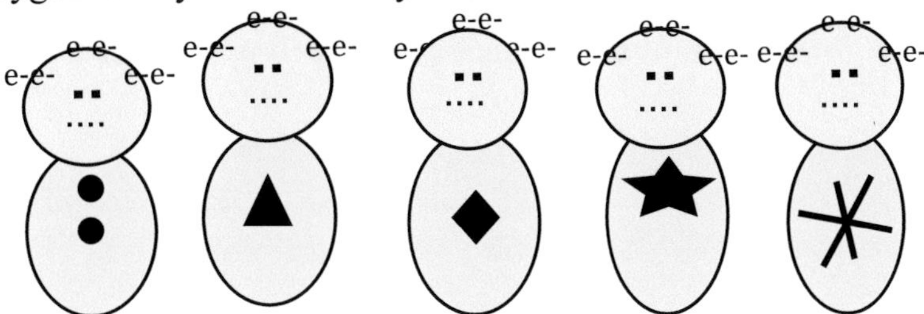

“We hope you never forget us Guy. These are the facts we want you to remember. Most important is the fact that we have 6 electrons in our Outside Energy Level. These are the electrons that are involved in forming compounds.”

CHEMICAL FAMILIES
GROUP 6A
THE OXYGEN FAMILY

In Group 6A each element has 6 electrons in its Outside Energy Level

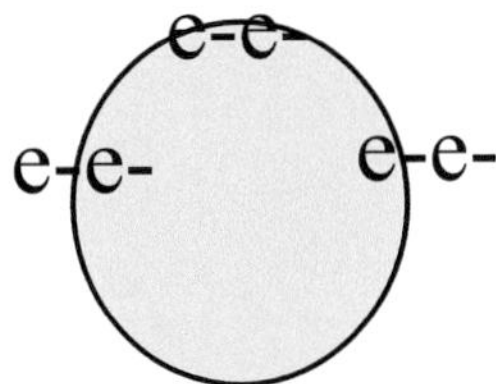

The first two elements in this family are similar acting as non-metals.
The last three are different. They are studied separately.

8	Oxygen	(O) makes up 20% of our atmosphere.
16	Sulfur	(S) is a soft yellow substance used in drugs.
34	Selenium	(Se) used in surge protectors to protect computers.
52	Tellurium	(Te) its garlic odor can identify this poison.
84	Polonium	(Po) is radioactive. So wear protective clothing.

Guy said, “I’ll remember all these facts about your family. I’ll remember all you, Oxygen family elements. I promise.”

THE HALOGEN FAMILY
Chapter 9

Professor Terry observed that Guy was enjoying meeting the Chemical Families and learning about elements in those families. She said to Guy, "Our trip has been a great success. I'm so glad you are meeting these elements now because you will see them again as you learn how these same elements combine to form compounds and engage in chemical reactions. When you meet them again, they will not just be a name to you. You will know them.

Professor Terry set the bubble down near the entrance to 7th Street. The beautiful light green balloons fell gently behind the bubble and deflated. Professor Terry moved them to a safe place and then they were ready to meet the elements on 7th Street.

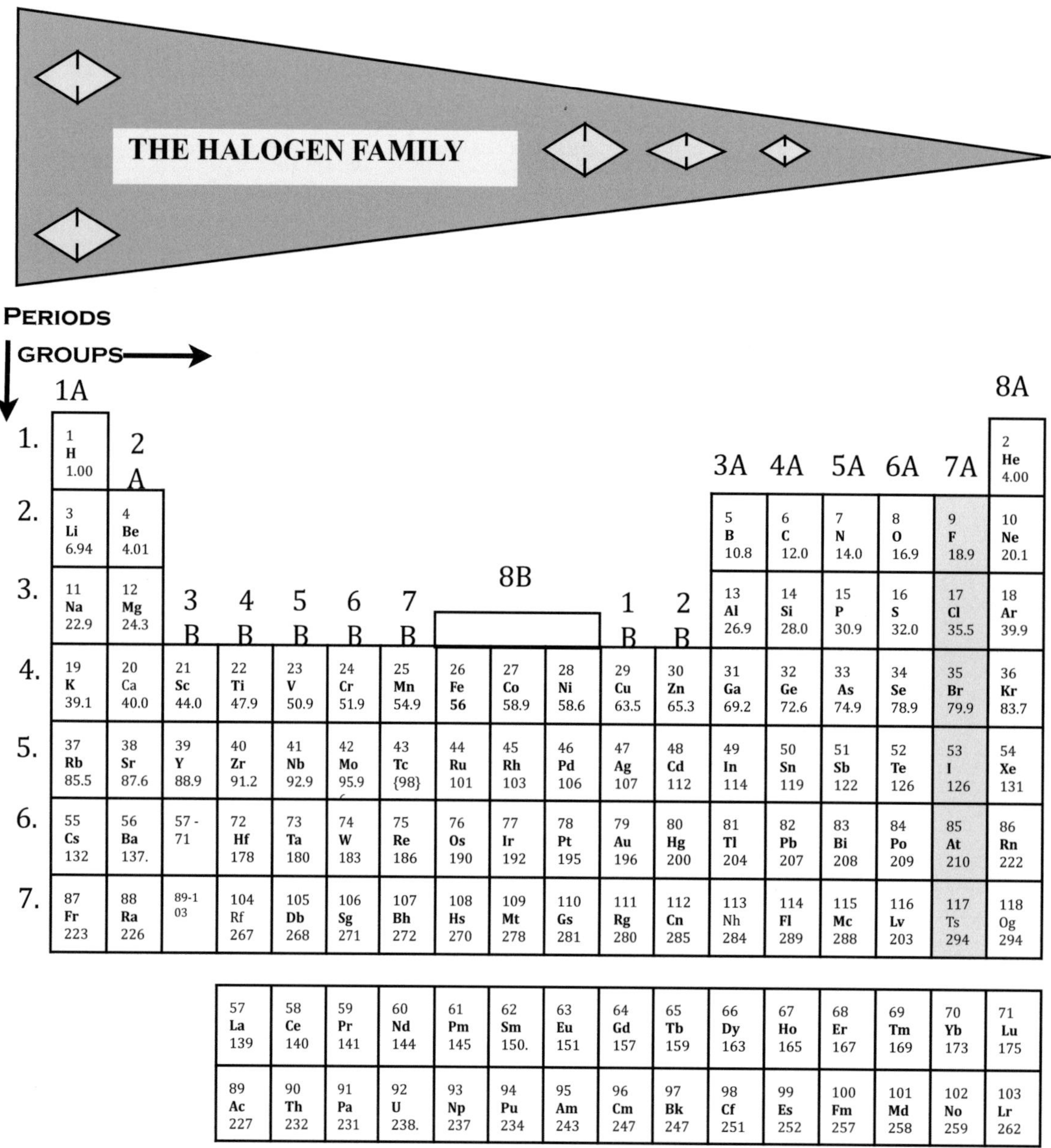

Periods	1A	2A	3B	4B	5B	6B	7B	8B	8B	8B	1B	2B	3A	4A	5A	6A	7A	8A
1.	1 H 1.00																	2 He 4.00
2.	3 Li 6.94	4 Be 4.01											5 B 10.8	6 C 12.0	7 N 14.0	8 O 16.9	9 F 18.9	10 Ne 20.1
3.	11 Na 22.9	12 Mg 24.3											13 Al 26.9	14 Si 28.0	15 P 30.9	16 S 32.0	17 Cl 35.5	18 Ar 39.9
4.	19 K 39.1	20 Ca 40.0	21 Sc 44.0	22 Ti 47.9	23 V 50.9	24 Cr 51.9	25 Mn 54.9	26 Fe 56	27 Co 58.9	28 Ni 58.6	29 Cu 63.5	30 Zn 65.3	31 Ga 69.2	32 Ge 72.6	33 As 74.9	34 Se 78.9	35 Br 79.9	36 Kr 83.7
5.	37 Rb 85.5	38 Sr 87.6	39 Y 88.9	40 Zr 91.2	41 Nb 92.9	42 Mo 95.9	43 Tc {98}	44 Ru 101	45 Rh 103	46 Pd 106	47 Ag 107	48 Cd 112	49 In 114	50 Sn 119	51 Sb 122	52 Te 126	53 I 126	54 Xe 131
6.	55 Cs 132	56 Ba 137.	57 - 71	72 Hf 178	73 Ta 180	74 W 183	75 Re 186	76 Os 190	77 Ir 192	78 Pt 195	79 Au 196	80 Hg 200	81 Tl 204	82 Pb 207	83 Bi 208	84 Po 209	85 At 210	86 Rn 222
7.	87 Fr 223	88 Ra 226	89-1 03	104 Rf 267	105 Db 268	106 Sg 271	107 Bh 272	108 Hs 270	109 Mt 278	110 Gs 281	111 Rg 280	112 Cn 285	113 Nh 284	114 Fl 289	115 Mc 288	116 Lv 203	117 Ts 294	118 Og 294

57 La 139	58 Ce 140	59 Pr 141	60 Nd 144	61 Pm 145	62 Sm 150.	63 Eu 151	64 Gd 157	65 Tb 159	66 Dy 163	67 Ho 165	68 Er 167	69 Tm 169	70 Yb 173	71 Lu 175
89 Ac 227	90 Th 232	91 Pa 231	92 U 238.	93 Np 237	94 Pu 234	95 Am 243	96 Cm 247	97 Bk 247	98 Cf 251	99 Es 252	100 Fm 257	101 Md 258	102 No 259	103 Lr 262

Guy Meets the Halogen Family 7th Street in Periodic Table Land

Guy said, "If my eyes were closed I'd know I was on 7th Street. There's that wonderful aroma of pine. Guy looked all around him and saw the magnificent pine forest ahead of him, and there was the entrance to the trail that led to 7th Street, the home of the Halogen family. Right above the pine forest was their family flag flapping in the breeze.

They started walking along the trail through the pine forest. Guy could not resist kicking the pine cones along the way. As they approached 7th Street, Professor Terry shared some information that she knew about the Halogens. "The name Halogens means *salt makers.* They form salts when they combine with a metal. When they combine with Hydrogen, they form acids. I see Chlorine ahead," observed Professor Terry. "He will tell us more about the Halogen family."

Chlorine was working on a rectangular sign on which he had inscribed the Halogen family name. "This is going at the top of our banner," said Chlorine proudly showing the Halogen sign to Professor Terry and Guy.

The Halogen Family

Chlorine put his work down and began talking with Professor Terry and Guy. "I overheard everything that you told Guy about this family, and your facts are correct. "I might add that the elements in our family represent all three states of matter. Fluorine and Chlorine are gases. Bromine is a liquid. Iodine and Astatine are solids. The elements will enjoy showing you their patches and telling you about their elements and compounds they form."

Then Guy wandered down to Fluorine's house to talk with him.

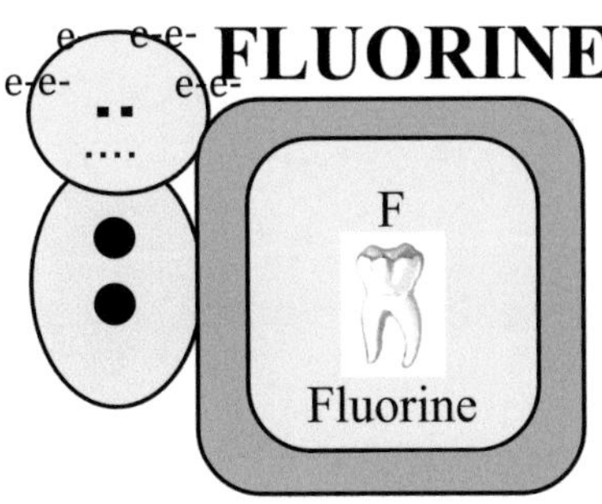

FLUORINE, "I'm a pale yellow gas, and the lightest member of this family. I'm *very* reactive. So, I'm usually found in the form of the solid mineral called Fluorite, a compound of Fluorine**.** I've been known since the 1500's to lower the melting points of metals when smelting ores. Fluorine's compounds are used in steel making, in refrigerants, in making teflon for frying pans, and in dentistry to prevent cavities. I'm drawing a tooth in my square."

Chlorine and Professor Terry caught up with Guy. Chlorine informed Guy, "I'm next. I'll tell you all about my element, the compounds I make, and uses I have."

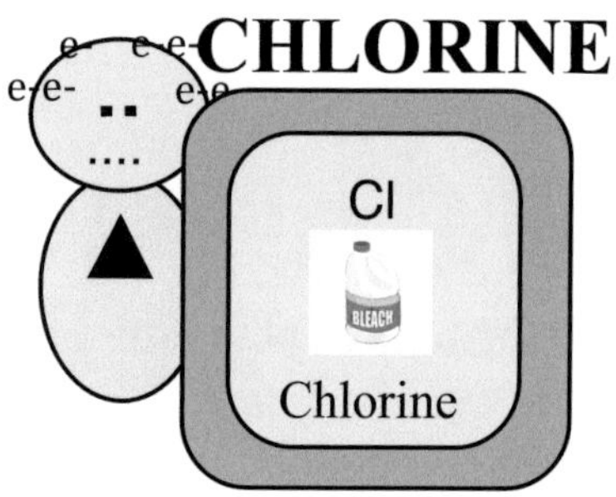

"I'm **CHLORINE,** a greenish yellow gas. I am so poisonous countries have agreed never to use chlorine again in war. As the salt, Sodium Chloride, I am used to make food taste good. I'm found in ocean water that covers three-fourths of the earth's surface. Where seas have dried up, I'm there in salt flats. I combine with Hydrogen to make hydrochloric acid. I also form several Carbon compounds: Chloroform used in operating rooms as anesthesia, and Carbon tetra chloride used in dry cleaning. Of my many household uses, the best known is bleaching clothes extra white. I'm also used to kill germs in water. I'm drawing a bottle of bleach for my square."

Chlorine chatted a while longer with Guy. "You have so far met the two elements that are gasses. Next you are going to meet Bromine who is a liquid. Do you know that there are only 2 elements that are liquids ? One of them is our own Bromine."

Guy said, "When I first heard that, I thought, *Wait a minute–the world is full of all different kinds of liquids. Not just two!* Then Professor Terry reminded me that she was talking about elements. The only two liquids she was talking about were elements. The rest of the liquids in the world were compounds or mixtures. Then it made more sense."

"Let's go, Guy, I will take you to visit Bromine. He will be happy to tell you all about his element."

"I'm **BROMINE,** a reddish brown liquid. I am found in brine pools in three places: in Israel's Dead Sea, in the state of Arkansas in the USA, and in China. My compound, silver bromide, is used in photography. I was used to make clothing fire retardant, but they found that I was too toxic. I'm unique for another reason. I'm one of only two elements in the whole universe that is a liquid. Mercury is the other element. Being a liquid makes me unique. So, I'm drawing an erlenmeyer flask filled with my reddish brown liquid for my square."

Chlorine said, "The next two elements we will visit are solids. One of these you know well. Wait and see!"

Guy said, "Let me guess. The element I will know is Iodine. My mother puts Iodine on my knees when I scrape them. But the Iodine she uses is a liquid."

Chlorine responded, "The Iodine you put on your cuts is Iodine dissolved in alcohol. One of the properties of Iodine is that it kills germs. Soon you're going to meet Iodine, the solid."

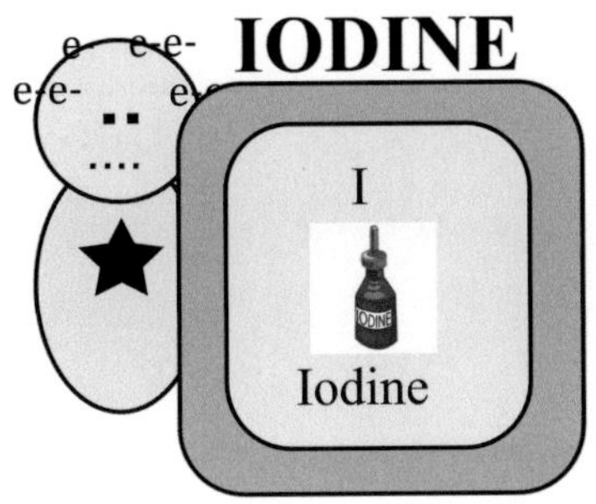

IODINE was out in front of his house drawing a bottle of iodine on his square. “I’m a bluish black solid. I’m a brown color when put in alcohol. I’m used to kill germs. Before surgery, surgeons paint me on the area of the body to be operated on.” Iodine continued, “If I stand too long I can sublimate, which means I can change from a blueish black solid into a violet gas without ever becoming a liquid. The word, Iodine, comes from a Greek word that means violet. Guy, your body’s thyroid gland needs Iodine to stay healthy. ”

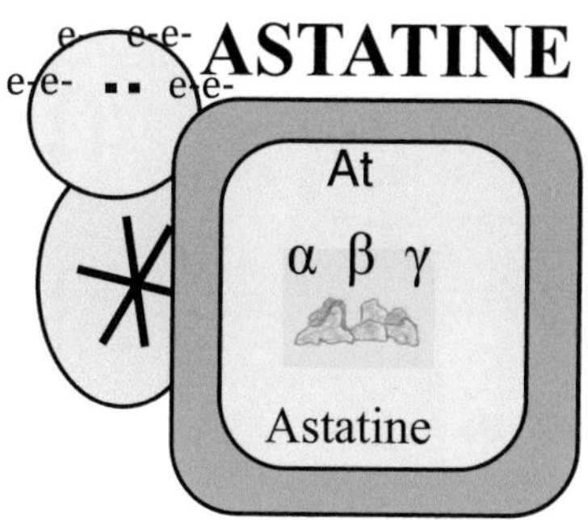

ASTATINE is a solid element that is radioactive. “Because I give off radiation, I have been used in small amounts in medicine to treat cancer.” Astatine said, “I’m just drawing rocks giving off radiation for my square. The Greek letters above my rocks stand for Alpha, Beta and Gamma radiation. I am essentially a non-metal. Rarely, I have only made compounds with Carbon, yet some people want me to be called a metalloid.”-T

Chlorine bringing the visit to an end said, “Element Tennessine-Ts, Atomic Number 117, is also a member of our family. He’s one of those elements created in a lab with a half life of seconds or less. So he has no uses. That’s why he doesn’t have a patch. He may not have any uses as yet, however, we are happy that scientists proved that this element exists.”

Suddenly, the five little Halogens came running over to see Professor Terry and Guy.

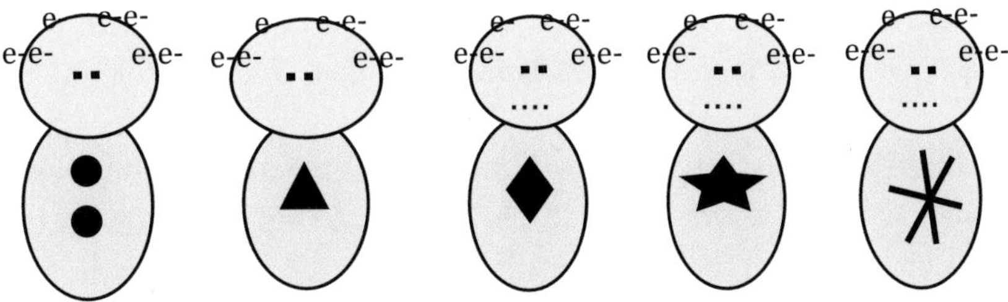

“Now that you have met all the members of our family we have a gift for you.” All the Halogens gathered around Guy to give him the index card with all the important information on it.

CHEMICAL FAMILIES
GROUP 7A
THE HALOGEN FAMILY

In Group 7A each element has 7 electrons in its Outside Energy Level

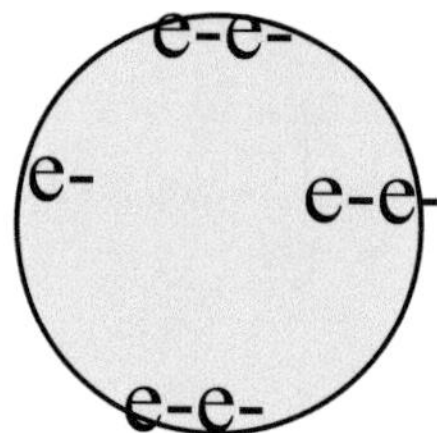

These elements are called Halides, salt makers. Like the Alkali Metals, they are extremely reactive and make compounds easily. They react mostly as non-metals.

9 Fluorine (F) used to make teeth stronger to prevent cavities.
17 Chlorine (Cl) purifies swimming pools.
35 Bromine (Br) I'm a reddish brown liquid element.
53 Iodine (I) keeps the thyroid gland functioning properly.
85 Astatine (At) used in cancer research.

One Halogen said, "The most important fact to remember is we have 7 electrons in our Outside Energy Level."

Another Halogen said, "Remember also that our family has elements representing all three States of Matter–solids, liquids and gasses.

Another said, "Best of all we are Halides—salt makers."

Chlorine said, "We don't want you to forget us, Guy."

Guy said, "How could I forget you with this nice card you gave me." Looking at the index card once more, Guy said, "As I look at your green circle with the 7 electrons around it, I'll always remember that the Halogens have 7 electrons in their Outside Energy Level."

Then he and Professor Terry jumped into the bubble and off they went to see the Noble Gas Family. It was a short. trip and Guy realized that his visits to the Chemical Families would soon be over. The Noble Gas Family was the last family in Periodic Table Land.

THE NOBLE GAS FAMILY

Chapter 10

Guy spied the flag of the Noble Gas Family in the distance. "I see the purple flag flying high over 8th Street. Looks like we'll be there in a few minutes."

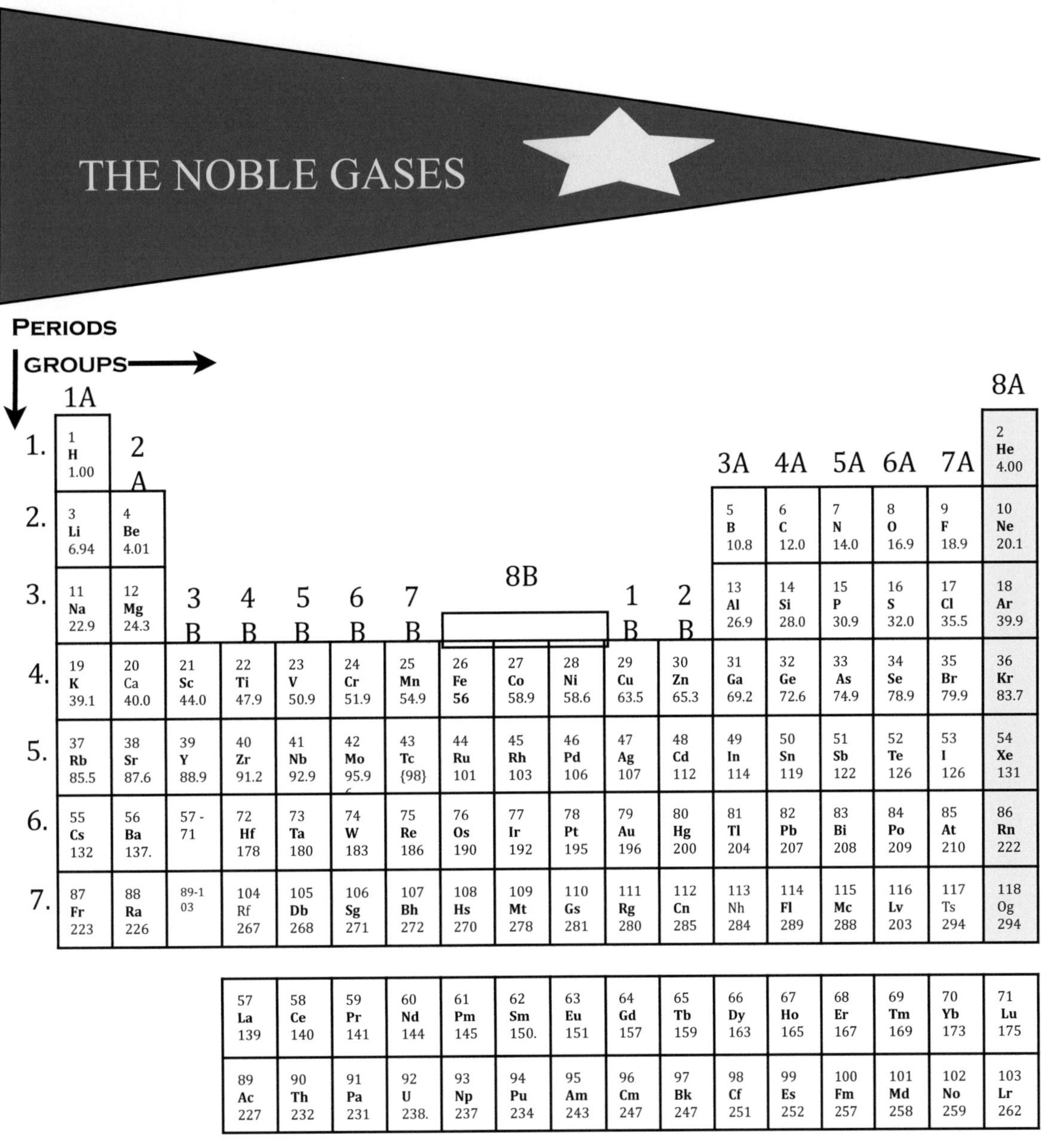

Period	1A	2A	3B	4B	5B	6B	7B	8B	8B	8B	1B	2B	3A	4A	5A	6A	7A	8A
1.	1 H 1.00																	2 He 4.00
2.	3 Li 6.94	4 Be 4.01											5 B 10.8	6 C 12.0	7 N 14.0	8 O 16.9	9 F 18.9	10 Ne 20.1
3.	11 Na 22.9	12 Mg 24.3											13 Al 26.9	14 Si 28.0	15 P 30.9	16 S 32.0	17 Cl 35.5	18 Ar 39.9
4.	19 K 39.1	20 Ca 40.0	21 Sc 44.0	22 Ti 47.9	23 V 50.9	24 Cr 51.9	25 Mn 54.9	26 Fe 56	27 Co 58.9	28 Ni 58.6	29 Cu 63.5	30 Zn 65.3	31 Ga 69.2	32 Ge 72.6	33 As 74.9	34 Se 78.9	35 Br 79.9	36 Kr 83.7
5.	37 Rb 85.5	38 Sr 87.6	39 Y 88.9	40 Zr 91.2	41 Nb 92.9	42 Mo 95.9	43 Tc {98}	44 Ru 101	45 Rh 103	46 Pd 106	47 Ag 107	48 Cd 112	49 In 114	50 Sn 119	51 Sb 122	52 Te 126	53 I 126	54 Xe 131
6.	55 Cs 132	56 Ba 137.	57 - 71	72 Hf 178	73 Ta 180	74 W 183	75 Re 186	76 Os 190	77 Ir 192	78 Pt 195	79 Au 196	80 Hg 200	81 Tl 204	82 Pb 207	83 Bi 208	84 Po 209	85 At 210	86 Rn 222
7.	87 Fr 223	88 Ra 226	89-103	104 Rf 267	105 Db 268	106 Sg 271	107 Bh 272	108 Hs 270	109 Mt 278	110 Gs 281	111 Rg 280	112 Cn 285	113 Nh 284	114 Fl 289	115 Mc 288	116 Lv 203	117 Ts 294	118 Og 294

57 La 139	58 Ce 140	59 Pr 141	60 Nd 144	61 Pm 145	62 Sm 150.	63 Eu 151	64 Gd 157	65 Tb 159	66 Dy 163	67 Ho 165	68 Er 167	69 Tm 169	70 Yb 173	71 Lu 175
89 Ac 227	90 Th 232	91 Pa 231	92 U 238.	93 Np 237	94 Pu 234	95 Am 243	96 Cm 247	97 Bk 247	98 Cf 251	99 Es 252	100 Fm 257	101 Md 258	102 No 259	103 Lr 262

"Oh look below. See that lake reflecting the green trees around it. The whole lake looks green. It's so beautiful! I keep seeing new parts of my magnificent world all the time. That makes me love my beautiful world more and more. I appreciate learning chemistry that reveals the world hidden below this world I love."

Guy Meets the Noble Gas Family
8th Street in Periodic Table Land

"Oh look! There's 8th Street. I see it clearly now." Guy was ready to meet the Noble Gases, the last family in Periodic Table Land. After they landed in the cleared area behind 8th Street, Professor Terry and Guy walked quietly through another pine forest not quite as big as the one on 7th Street. They stopped at the entrance to 8th Street to look at the family flag. There it was—a beautiful purple silk flag decorated with a golden star and their family name, The Noble Gases. It was flying straight out as the wind had developed into a stiff breeze. Admiring the flag, Guy turned to Professor Terry and exclaimed, "The flag certainly looks regal, very appropriate for elements so noble. Why do they call the elements on 8th Street Noble Gases?"

Professor Terry said, "It's because these elements do not like to mingle with the other elements and form compounds. Like members of royal families, they keep to themselves. They used to call these elements ***Inert*** because they **never** formed compounds. Once scientists were able to take some of these elements and create a few compounds in the lab, their name was changed to **Noble**. Just so you know Guy, Argon, Krypton, and Radon have formed compounds with Oxygen and Fluorine. In general these elements form very few compounds but you can not call them inert which means not able to form compounds."

"This gives you a little background about the Noble Gas Family. Now let's go meet these elements that are noble." They started down 8th Street and noticed it was painted gold like the star on their flag. Purple is too dark to see writing through the color, so they used gold, the next most regal color. Just ahead was #2, 8th Street the home of Helium.

Helium was not far away drawing balloons on his festival patch. The real balloons were nearby tied to the trees. He was trying to draw the balloons to look as realistic as possible. Helium looked up and saw his two visitors standing there. He put aside his paint brush and was anxious to tell them about himself and his festival patch.

Helium said, "I heard Professor Terry's explanation about why our family is called Noble, and it was accurate. So I don't have to repeat it but there's one more thing for you to notice. We live on 8th Street, and so we all should have 8 electrons in our Outside Energy Level and 8 e^- hair. Notice I have only 2 electrons and 2 e hair. I have to tell you why I was put into Group 8A with the Noble Gases and not in Group 2A where all the elements have 2 electrons in their Outside Energy Level and 2 e^- hair like me. The reason is my 2 electrons make my Outside Energy Level complete because I have only one energy level. The elements in Group 2A need 6 more electrons to be complete. So I'm more like these Noble Gases whose Outside Energy Levels are complete. Now you're ready to meet the elements in the Noble Gas Family."

Helium said, "First I'll tell you about my element and then you can visit the rest of the family. Come on over and sit down on my bench under this Linden tree. It's nice and cool here."

Here's the little piece I've made to put at the top of our family banner. I thought I'd show you this first and then my own patch."

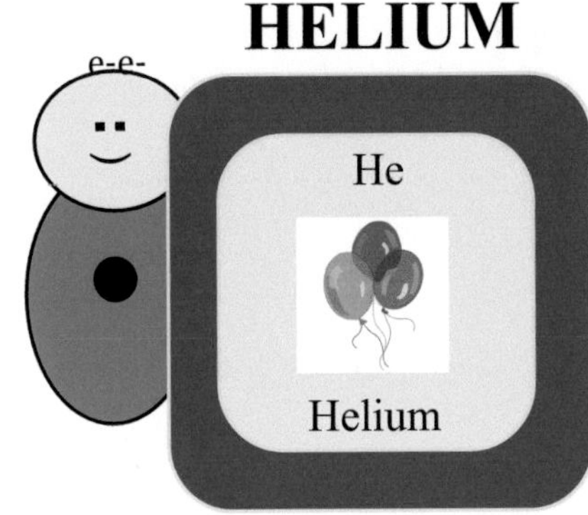

"I'm **HELIUM,** the lightest of all the Noble Gases. I'm so light that I escape the earth's atmosphere, and go off into space. I'm the second most abundant element in the universe. I am found in the sun and stars. On earth, they use me to inflate blimps and party balloons. I do not have any compounds under normal conditions." Guy thanked Helium and wandered down 8th street to the home of Neon.

NEON was next. As Guy passed by the house, he peeked into his side window and saw many different colored signs decorating the inside of his home. It was so pretty. Neon came out and told Guy, "Those famous bright lights of Broadway in New York City are Neon signs**.** That's what made me famous." Neon said, "I'm used as lasers, in diving equipment, switching gears, and lightning rods. I'm drawing a Neon sign on my square that could be put in a shop's window."

Guy moved on to see **ARGON**. "I make up 1% of the earth's atmosphere," said Argon. "Scientists can take me out of the air by doing a fractional distillation of liquid air. This is a special process. I'm used in double pane windows, because I prevent heat transfer through the window. I'm drawing a light bulb because I make them last longer."

"I'm **KRYPTON**, and I'm famous because I can drain Superman of all his power. That's only true in comic books. I make compounds with Fluorine, Oxygen and Nitrogen. Krypton Fluoride is my most important compound. I'm used in high powered lasers and high speed flash photography. I'm drawing a tape measure displaying 1 meter because my wavelength on the electromagnetic spectrum is used to define the official length of the meter."

"I'm **Xenon**, and I've got a lot of uses. I'm used in headlights of new cars. I'm used in arc lamps, lasers, flash lamps, and as an ion propellant in spacecraft. I'm even used in anesthesia in hospitals. I'm going to draw a spaceship for my square."

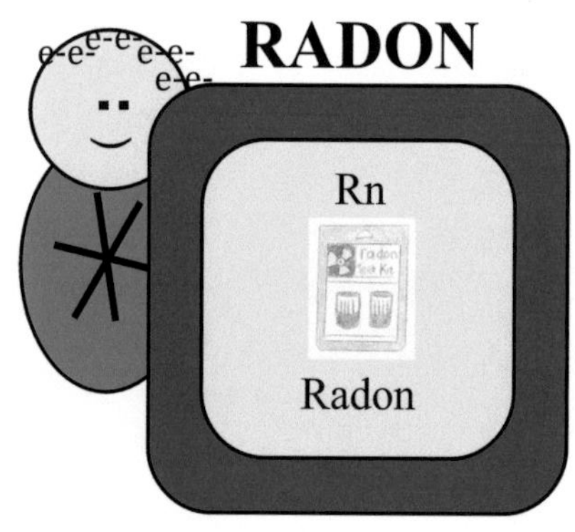

RADON said, "My atom is a radioactive gas that is so intensely radioactive that it prevents scientists from studying it. It's formed from the radioactive decay of Uranium. Uranium turns into Radium which in turn turns into Radon. It will also be around for tens of thousands of years. It seeps up through cracks in the earth and can get into houses to cause lung cancer. So air out your basement. My square will show a Radon Gas Detector.

"There's one more element that has been discovered in our family. It is element 118. It is extremely radioactive and has been named Oganesson-Og. The atom was created in a laboratory, but it stayed in existence for less that a millisecond. So, it's no surprise that it doesn't have any uses."

Just as they were about to leave, Helium, Neon, Argon, Krypton, Xenon, and Radon came running after them with an index card containing all the Noble Gas Family facts."

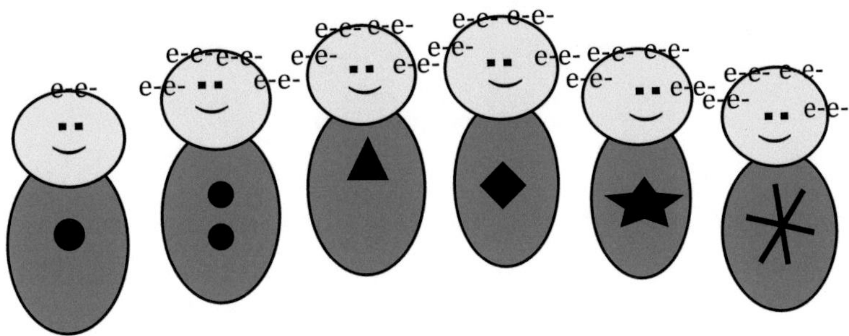

They were calling, "You can't leave until we give you our family fact card. We don't want you to forget what you have learned about us. Here it is now. This is our farewell gift to you."

The card was neatly done. The little atoms that created it are proud of their work, as they should be. They made it carefully. "We all love you Guy," they said in unison.

Guy said, "You are all so very kind. I've enjoyed meeting you."

CHEMICAL FAMILIES
GROUP 8A
THE NOBLE GAS FAMILY

In Group 8A the elements have 8 electrons in the Outside Energy Level,

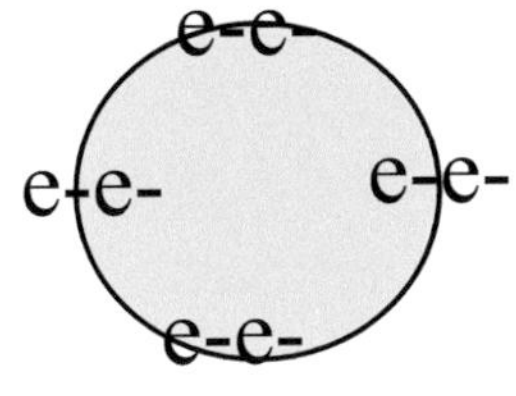

Exception

Helium\ has only 2 electrons
because he only has one Energy Level.

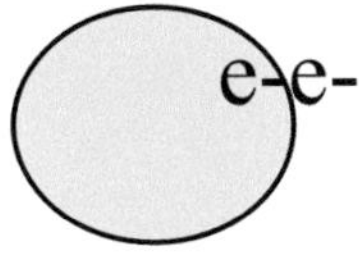

2 Helium (He) is used in balloons and has only 2 electrons.
10 Neon (Ne) is used in advertising signs.
18 Argon (Ar) is used in light bulbs to prolong their life.
36 Krypton (Kr) is special because its wavelength is exactly 1 meter long.
54 Xenon (Xe) is used in the headlights of new cars.
86 Radon (Ra) is a radioactive health hazard——air out your basement.

Helium said, "There's one thing before you go. You must not forget how many electrons we have in our Outside Energy Level. That's what our family fact chart tell you. All of us have 8 electrons there except Helium who has 2. The part to remember is that all of us including Helium have complete Outside Energy Levels."

Guy looked over the index card, and he was about to go when shy little Xenon said, " I'm going to give you a reason to stay here a little longer. The judges have finished picking the winners of the flag contest. Come with us over to Town Hall so we can see how the contest turned out."

****

Over at Town Hall, all the Chemical Families gathered around in the courtyard waiting for the results of the flag contest. Professor Terry and Guy joined the Noble Gases.

It was a busy scene around Town Hall. You could feel the excitement in the air. The results were about to be announced. There was lots of chattering and enthusiastic predictions. The children were running around and playing tag. Each of the families anticipated being the winners. Then the head judge stepped on to the porch with the winner sheet in his hands. Everyone quieted and formed a semicircle in front of Town Hall. The judge began making his announcement.

"I will start with last place and work my way up to the Grand Prize winner. However, before I begin I must tell you that you are all winners. You all have beautiful flags and banners to fly at the head of your street from now on whether you are winners or not. The flags and banners are all lovely. It was difficult to pick the winners. However, the job is done and here are the winners..........."

The winners were named, and when he was finished, he posted the list on the pole in the center of the courtyard to further honor the winners. "Let's wander over and look at the list," said Xenon. Guy and Professor Terry joined Xenon and went over to the pole where the winner's list was posted. They wanted to make sure to remember the winners correctly.

RESULTS OF THE FLAG CONTEST

Best in Show	- The Noble Gases, Group 8A
Most Meaningful	- The Nitrogen Family, Group 5A
Most Beautiful	- The Oxygen Family, Grou[6A
Honorable Mention	- The Alkaline Earth Metals, Group 2A

"Guy, this brings our visits to the Chemical Families to an end. You have met the members of each of the families and gotten to know a little about each of them. Now it's time to return to the lab," announced Professor Terry. Seeing that Guy was sad to have his visits with the Chemical Families end, she then reassured Guy that there were many more adventures ahead.

"Next, Guy, you are going to find out how the elements in these families join together to form something that is new and exciting—compounds. That will be your next adventure in your quest to discover more about your magnificent world. It's the story of how the elements become Happy Atoms. It will be your best adventure yet. So I know you will smile when when you see the elements become Happy Atoms. However, before your next adventure, Sodium has an investigation to do. He will not need our help. So you can have some time off to enjoy part of your summer vacation with your family up on the mountain. I will follow what Sodium is doing and let you know when it's time to go back to Periodic Table Land and begin your new adventure."

****

Back at the cabin that evening, Guy looked out the large picture window in his room with its beautiful view of the stars and the moon. Guy saw Wish Star up in the sky smiling down on him and waving his magic wand.

Guy smiled back knowing Wish Star was there for him always. As he closed his eyes and started falling asleep, thoughts of this new adventure bounced around in his head. He thought of all the little elements he had met. He now knew each one was part of a family. They had special colors, lovely flags flying at the entrance to their special streets. He had learned so much about each of the elements in these families. These elements were going to be part of this next adventure. He tossed around all sorts of possibilities. Then looking up to take one more glimpse of Wish Star, he fell asleep thinking. "What will my next adventure be like? "Guy couldn't wait to find out.

BOOK 2
PART3
SODIUM'S SEARCH
FOR
HAPPY ATOMS

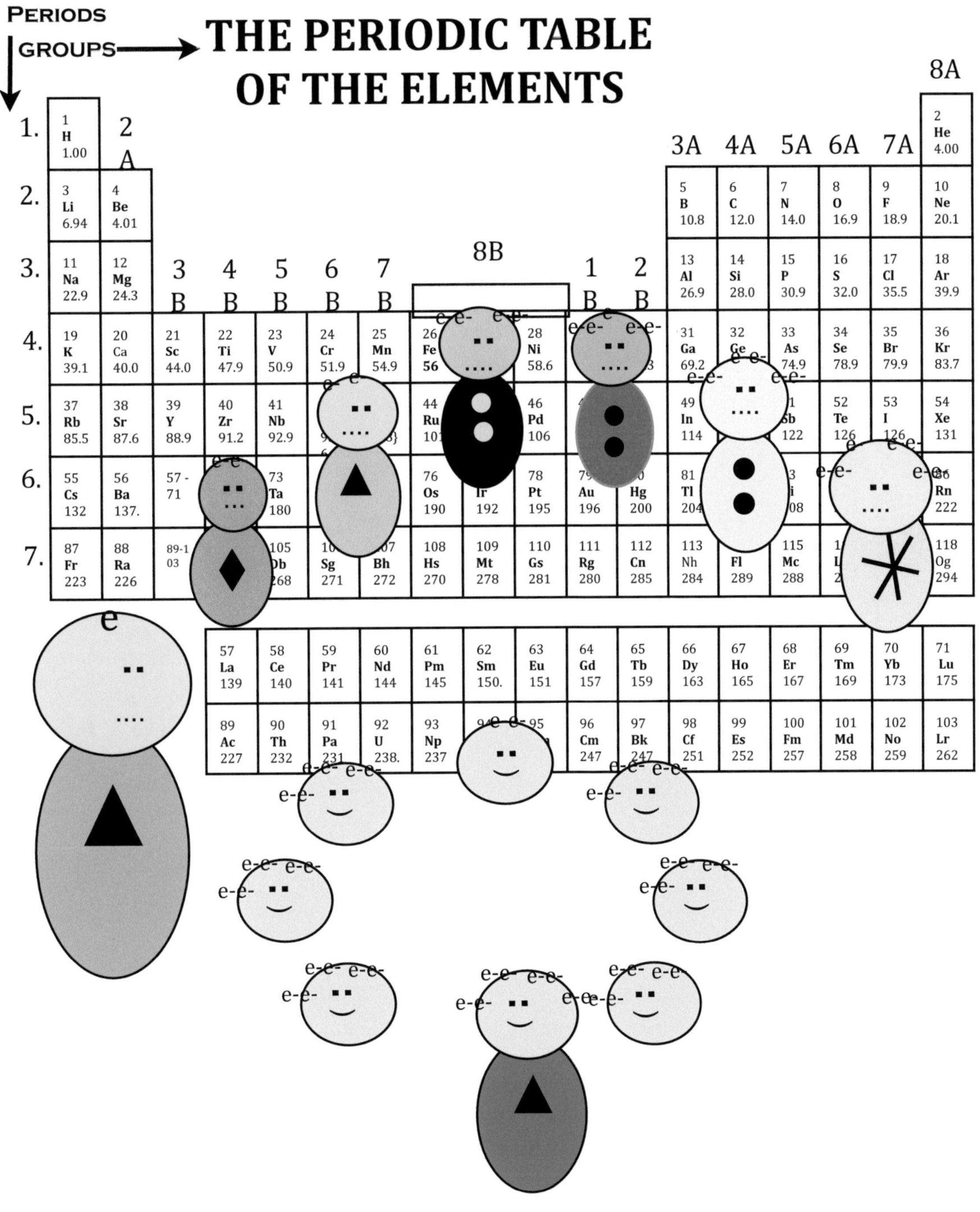

BOOK 2

Part 3

Sodium's Adventure

TABLE OF CONTENTS

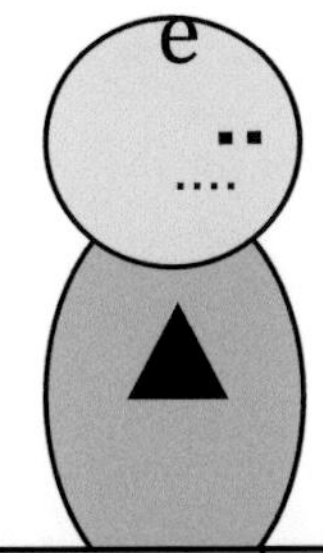

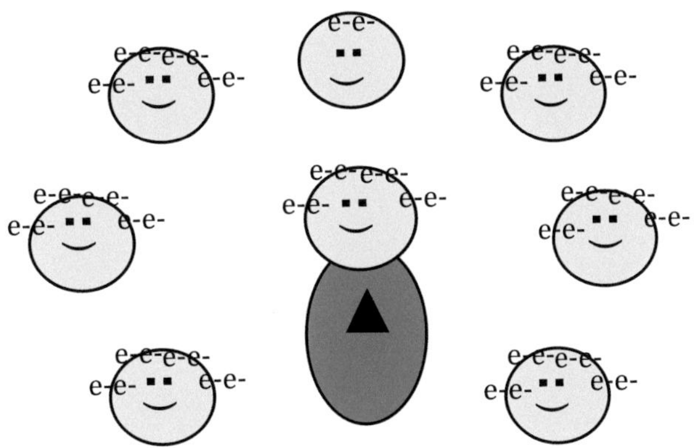

The Alkali Metals Send Sodium on a Mission
Chapter 1

Sodium had become good friends with Professor Terry and Guy, visiting them often. Professor Terry noticed that Sodium was looking sadder and sadder each time he visited. Good friends instinctively know when a friend has a problem, and she felt this about Sodium. One day when she had a chance to talk with Sodium privately, she said, "I feel something is troubling you, Sodium."

Sodium needed to share his problem and so revealed to her that deep down his whole family had been extremely sad. "We have put on a good face for a long time; but when my family is alone, tears are running down our dear little atom faces. When I see my family so sad, it breaks my heart. I just wish we were all happy, but the fact is we're not."

Professor Terry's response was, "Sodium, I think you need to get to the bottom of this situation. Go back to your Alkali Metal Family and discuss the problem with them."

Sodium considered her advice and said, "We are definitely tired of being sad all the time. It seems we've been this way forever, but it is getting worse. I guess it's time we do something about it. I'll go talk to my family."

"Good luck, Sodium! I don't want my good friend to be so unhappy. Maybe your family will come up with a good suggestion that will get you out of this predicament. I have found that when groups come together to solve problems, good things happen," advised Professor Terry.

Sodium returned to First Street. All the little Alkali metals came together, tears running down their dear little atom faces. It was definitely time to do something about this. Sodium was resolved to find a solution to this problem. So, Sodium arranged for a family meeting that very same day.

They all gathered together wondering what Sodium was going to say.

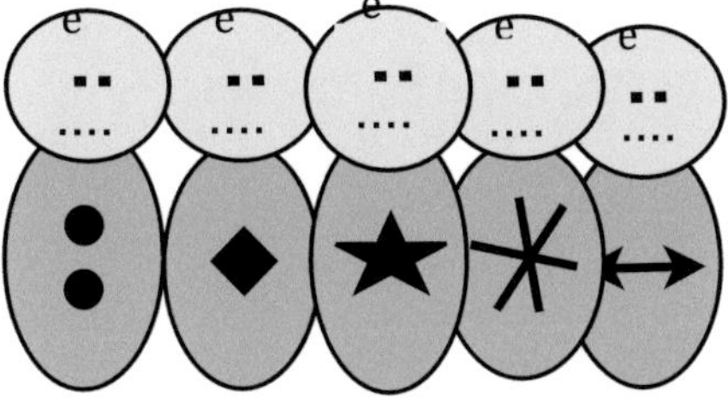

In the most positive tone of voice he could muster, Sodium spoke up. "Look at us. Look at all our tear stained faces. We should not be this sad. We need to find a way to be Happy Atoms. We need to talk about how we can do this?"

Thoughtfully, Lithium, Li said,

"I'd like to know the answer to that question, but I have no ideas."

Rubidium, Rb with his face twisted from crying said,

"Do you think we are the only sad family in Periodic Table Land?"

They all started to talk among themselves trying to figure out what they could do to become Happy Atoms.

After sitting and pondering all the suggestions, Potassium, Cesium and Francium had an idea. They finally spoke up and shared it with the rest of the family,

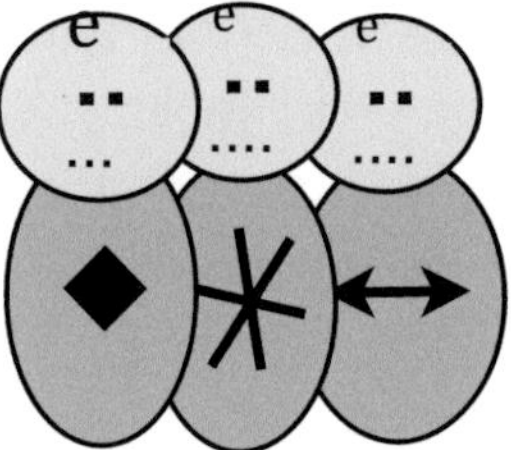

"What we have decided is that an investigation is in order. We need to get to the bottom of this sad situation. We need to find out if anyone else in Periodic Table Land is as sad as we are. Maybe some element out there has an idea of how to become happy. None of us can figure it out."

Sodium, standing up as tall as he could, summarized what he heard his family suggesting. "I hear a plan. We will search for and talk to both sad and happy atoms, and we will do this as scientifically as possible. We will gather and record as much data as we can as we speak with these sad and happy atoms. As a result of our research, hopefully, we'll find out how to become Happy Atoms."

All the elements in the Alkali Metal family elected Sodium to do this scientific research. They suggested that Sodium search Periodic Table Land, street by street, to discover how they could become Happy Atoms."

Then the elements went off to the family garage and readied their lovely green balloons for Sodium. He slipped into the bubble below the inflated balloons, pointed his magic wand upward, and off he went. His family called after him,

"Good luck Sodium!"said Cesium.

"We're counting on you," said Potassium.

Lithium waved a sad farewell.

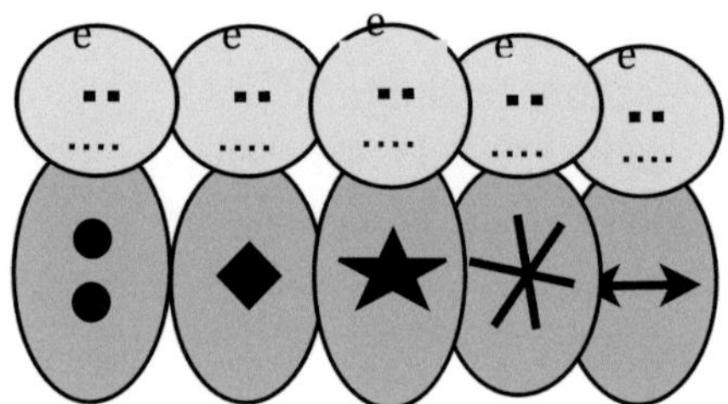

Then they all called after him,"God speed, Sodium!"as the lovely green balloons with Sodium aboard slipped out of sight. Will Sodium learn how they can become Happy Atoms? They sure hoped he would. The happiness of their family hung in the balance.

Sodium Searches for Happy Atoms

Chapter 2

The light green Alkali Metal Family balloons rose high in the cloudless sky. Sodium looked down from his vantage point in the bubble and had a grand view of Periodic Table Land. It was just gorgeous. We definitely should not be sad in such a wonderful land. He was amazed at the beauty of the architecture. What Sodium liked best was all the beautiful, colorful gardens surrounding so many homes. Then, there was that splash of color that the Chemical Family flags lent to the wooded areas at the head of most streets. Sodium was headed toward 2nd Street, his first stop on his investigative tour of Periodic Table Land. As he circled the clearing in the woods at the head of 2nd Street, Sodium lowered the magic wand, and the balloons landed softly on the air strip behind 2nd Street. As Sodium slid out of the bubble, he wondered what he would discover when he met the Alkaline Earth Metals. He planned to first visit his old friend Calcium. As he walked down 2nd Street, he passed the home of Beryllium. There was an open window, and Sodium thought he heard a gentle sobbing as he passed by.

Finally, he arrived at Calcium's home, #20, 2nd Street. After a short visit Calcium confirmed that his family was as sad as the Alkali Metals, and they had no idea how to become happy. Sodium told them that he was searching for a Happy Atom, hoping that he would find a way to help his family become happy.

Calcium, the leader of the Alkaline Earth Metal group said, "If you find out how to be happy, don't forget to tell us. We have no idea how to become happy, and we're as tired of being sad as you are." Calcium wished Sodium well, and off he went.

Soon the green balloons had Sodium flying high towards B Avenue. As Sodium entered the air space over B Avenue, his magic wand brought the balloons to a lower altitude. Flying low, he heard the Transition Metals crying uncontrollably. Their metallic houses amplified the noise of their misery. Sodium's ears hurt. He increased his speed as he couldn't get past B Avenue fast enough. The sobbing and wailing of all the Transition Metals on B Avenue was too much for his ears to endure. I'm not landing here. If they knew how to become Happy Atoms, they wouldn't sound like this. "Oh, will I ever get through here?" Finally, the green balloons broke into the silence past B Avenue. What a relief it was to get to 3rd Street.

Pointing his magic wand down, the green balloons soon landed Sodium's bubble on Boron's side lawn. He was the head of the Boron family. Aluminum was visiting Boron, and they invited Sodium to sit on the picnic bench with them. Boron offered Sodium some iced tea. Aluminum was talkative. It did not take long for Boron and Aluminum to share that 3rd Street was a sad place too.

Again Sodium was disappointed. He began to worry. He was halfway across Periodic Table Land now, and he had not yet found one happy atom. "My family will not understand if I come home without finding any happy atoms, or any idea about how we can become happy."

Sodium slipped into the bubble and was off to visit the Carbon family on 4th Street. They were also sad. Tears were running down their little atom faces. He looked at them and thought, "If I ever find out how to be a Happy Atom, there will be a lot of families in Periodic Table Land who will benefit from my discovery."

Again, the green balloons delivered Sodium safely to 5th Street. Here, he found the Nitrogen family as sad as all the other Chemical Families. He was really losing hope. He thought aloud, "I'm never going to find a Happy Atom, and so far no one even has an idea about how to become happy. At least my family had the sense to begin an investigation to search for a way to become happy." After visiting with the Nitrogen Family on 5th Street, he forced himself to push on and speak with the Oxygen Family on 6th Street. At this point he was totally discouraged.

The beautiful green Alkali Metal family balloons sailed once again high up into the cloudless sky. It gave Sodium a few minutes to think over what he was doing. He was well beyond the halfway mark in his investigation; and he knew no more now than when he started, except that his family was not alone in being so sad. Now he was at 6th Street.

Sodium landed behind Oxygen's house. Sodium always enjoyed visiting Oxygen because the air around Oxygen's home always smelled so fresh. Oxygen had an extra supply of that good fresh air that our lungs always appreciate. This time, Sodium was visiting Oxygen on a serious mission. He thought, "After this there will be little hope of finding a happy atom if I don't find one here."

Oxygen was out in front of his house. When he heard the balloons land, he ran around the house just in time to see Sodium sliding out of the bubble. Sodium gave Oxygen a friendly greeting, and they sat down at the picnic table to talk.

"Oxygen," said Sodium, "I'm here hoping that you might help me."

Oxygen looked into Sodium's troubled eyes and replied," You look stressed Sodium. How can I help?"

Sodium confided, "It all began because my family has been sad for so long. My job is to scientifically search Periodic Table Land to hopefully find a Happy Atom. In addition, I hope to find out how we all can become Happy Atoms. Well, I haven't found even one Happy Atom yet. I'm hoping your family might be happy and solve the whole problem."

Oxygen hung his head and thought for a minute about how he could tell Sodium that his family was not happy either. "I am so sorry, Sodium. I would like to be the one to help your family. However, we are as sad as all the rest of the families you have visited."

Sodium shared with Oxygen, "It's the same story all over. Now I find out that your Oxygen Family members are as sad as all the other elements in the rest of Periodic Table Land. How will I ever find out how to be a Happy Atom?"

This made Sodium almost turn back. Seeing all the elements so sad, he knew he had to continue searching for a solution. He had promised to search ***all*** of Periodic Table Land. There were now too many families depending on him. Only two more streets to go, and he will have searched all of Periodic Table Land. Sodium made a pledge aloud. even though there was no one around him at the time. "I will do my very best. As much as I have no hope at all any more, I must not quit until I have accomplished my mission. I said I would search *all* of Periodic Table Land, and I will. I'm a man of my word."

Sodium continued talking to himself out loud. "If I am going to behave as a true scientist, I must remain determined to go on in spite of the appearance of failure. I remember reading about Edison's experiments trying to create a light bulb. His experiments kept failing. The newspapers called him a fool and were relentless in making fun of him. In spite of their criticism, he kept trying. Finally, he created the light bulb and look how wonderful it has made life today. I have it much easier than Edison. No one is laughing at me. I will go on. There are only two streets left to search. I can make it! I will not turn back now."

The green balloon rose high in the sky again, and before long he was at 7th Street, the home of the Halogens. Sodium met with Fluorine, Chlorine, Bromine, Iodine and Astatine in hope that one of them would be a Happy Atom, To his dismay, all the Halogens were trapped in the gloom that hung over all the other elements in Periodic Table Land. Again, tears were running down all their little atom faces. Before he left, he talked with his friend Chlorine one last time.

Chlorine admonished him to make sure he spoke to Astatine before leaving. Chlorine said,"Astatine was talking about something the other day. We weren't paying him much attention. but I think it might be important for you to see what he as to say."

Sodium decided to talk to Astatine. Tears fell from Astatine's eyes as he recounted what he had heard, "Once when the wind was blowing from the direction of 8th Street I think I heard the sound of laughter. It was only once, but maybe there is some happiness over there."

Sodium was afraid to get his hopes up too high, but a ray of optimism crept into his spirit. Sodium thanked Astatine, got up, and left the Halogens. He headed back to where he had parked his green family balloons, slid into the bubble, and pointed his magic wand toward 8th Street. He dialed slow on the wand, wanting to delay the end of his investigation that seemed to be doomed to failure. 8th Street was his last chance for finding a Happy Atom in Periodic Table Land.

Astatine's words kept playing over and over in his mind—"I thought I heard a sound of laughter." These words kept his spirits up as he moved on to 8th Street. His family, and it appears, every other family in Periodic Table Land, depended on him. They deserved a chance to be happy. He had to keep trying."

The green balloons floated Sodium in his bubble slower than ever. Finally as he got nearer to 8th Street. Sodium began to hear what he thought were happy sounds. "It's probably my imagination," said Sodium out loud. "I am so desperate to find a Happy Atom somewhere that these sounds appear to be real." As the balloons moved Sodium even closer to 8th Street, these happy sounds became louder.

He landed his bubble in the clearing behind the wooded area at the top of 8th Street. He heard sounds that he had not heard in all of Periodic Table Land. There was definitely laughter. Sodium was excited and said, "This is not my imagination. I am really hearing laughter. I think I've found Happy Atoms."

Approaching the entrance to 8th Street, he even heard music, singing, and cheerful chattering. Suddenly he was flooded with hope. He knew that he had found what he had been searching for. 8th Street was definitely a happy place. Happy Atoms At Last!

Sodium Discovers What Makes Atoms Happy
Chapter 3

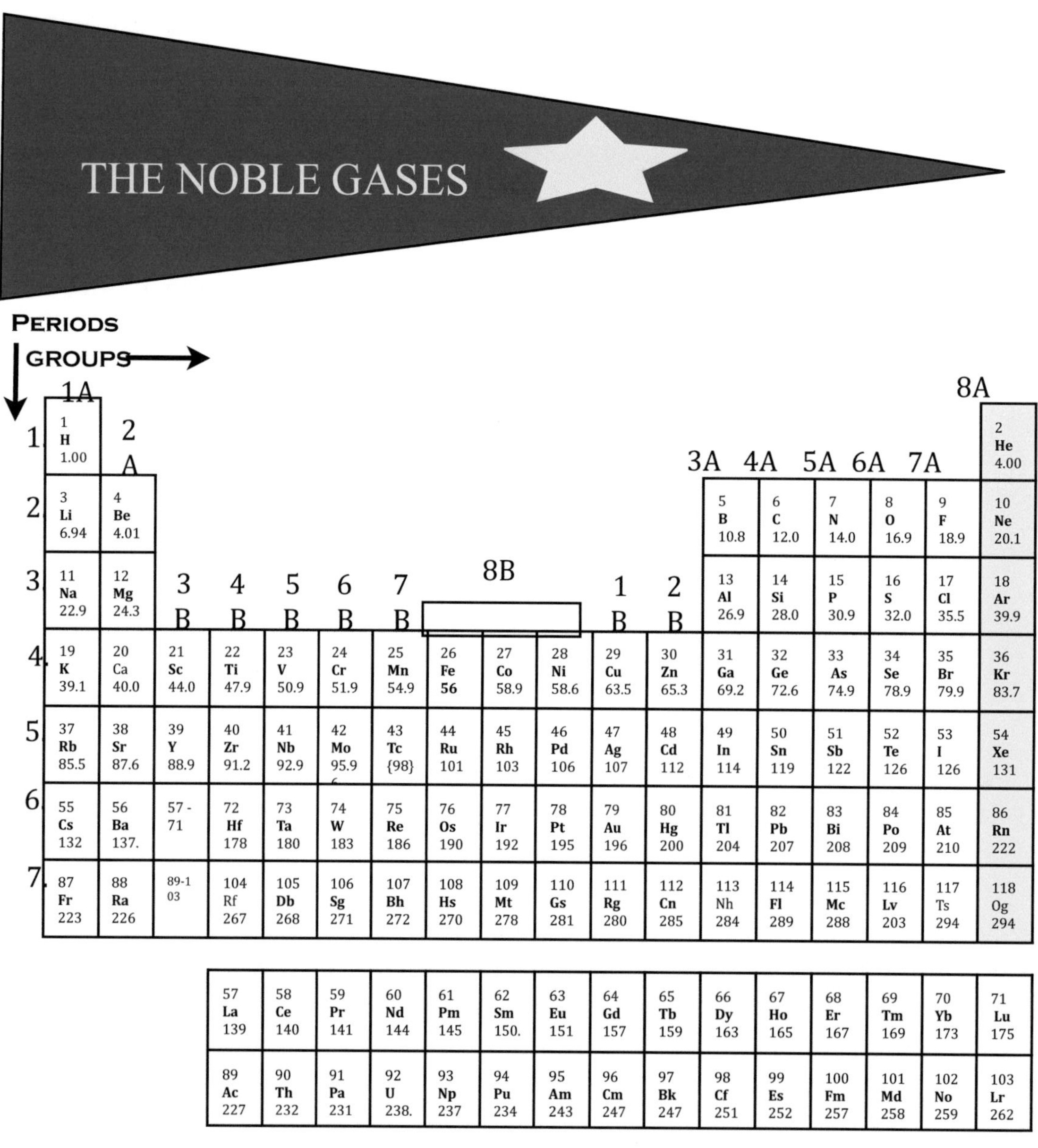

	1A	2A	3B	4B	5B	6B	7B	8B	8B	8B	1B	2B	3A	4A	5A	6A	7A	8A
1	1 H 1.00																	2 He 4.00
2	3 Li 6.94	4 Be 4.01											5 B 10.8	6 C 12.0	7 N 14.0	8 O 16.9	9 F 18.9	10 Ne 20.1
3	11 Na 22.9	12 Mg 24.3											13 Al 26.9	14 Si 28.0	15 P 30.9	16 S 32.0	17 Cl 35.5	18 Ar 39.9
4	19 K 39.1	20 Ca 40.0	21 Sc 44.0	22 Ti 47.9	23 V 50.9	24 Cr 51.9	25 Mn 54.9	26 Fe 56	27 Co 58.9	28 Ni 58.6	29 Cu 63.5	30 Zn 65.3	31 Ga 69.2	32 Ge 72.6	33 As 74.9	34 Se 78.9	35 Br 79.9	36 Kr 83.7
5	37 Rb 85.5	38 Sr 87.6	39 Y 88.9	40 Zr 91.2	41 Nb 92.9	42 Mo 95.9	43 Tc {98}	44 Ru 101	45 Rh 103	46 Pd 106	47 Ag 107	48 Cd 112	49 In 114	50 Sn 119	51 Sb 122	52 Te 126	53 I 126	54 Xe 131
6	55 Cs 132	56 Ba 137.	57 - 71	72 Hf 178	73 Ta 180	74 W 183	75 Re 186	76 Os 190	77 Ir 192	78 Pt 195	79 Au 196	80 Hg 200	81 Tl 204	82 Pb 207	83 Bi 208	84 Po 209	85 At 210	86 Rn 222
7	87 Fr 223	88 Ra 226	89-1 03	104 Rf 267	105 Db 268	106 Sg 271	107 Bh 272	108 Hs 270	109 Mt 278	110 Gs 281	111 Rg 280	112 Cn 285	113 Nh 284	114 Fl 289	115 Mc 288	116 Lv 203	117 Ts 294	118 Og 294

57 La 139	58 Ce 140	59 Pr 141	60 Nd 144	61 Pm 145	62 Sm 150.	63 Eu 151	64 Gd 157	65 Tb 159	66 Dy 163	67 Ho 165	68 Er 167	69 Tm 169	70 Yb 173	71 Lu 175
89 Ac 227	90 Th 232	91 Pa 231	92 U 238.	93 Np 237	94 Pu 234	95 Am 243	96 Cm 247	97 Bk 247	98 Cf 251	99 Es 252	100 Fm 257	101 Md 258	102 No 259	103 Lr 262

At the entrance to 8th Street, Sodium rested for a few minutes, thinking of what he would do next. The Noble Gases' royal purple flag above him was flapping gently in the breeze. Looking around, Sodium observed a bustling, lively scene. In the middle of 8th Street there was a three piece band. The smiling happy musicians were performing under a lovely Linden tree. The wide reaching branches provided cool shade for the Happy Atoms sitting beneath. The Noble Gas elements were singing, clapping, and obviously enjoying the music. Other elements were playing games. They were happy! There were little groups sitting around on picnic blankets talking and laughing. Sodium entered slowly and made his way down the street hidden partially by the bushes surrounding 8th Street. He was trying to not be too obvious, as he observed the activities all around him. He was overcome with awe, as he took in the jubilant scene. He couldn't believe there could possibly be this much happiness around him. His ears strained to absorb the laughter and music that filled the air everywhere. All the little atoms were as happy as happy can be. There was not a sad atom anywhere.

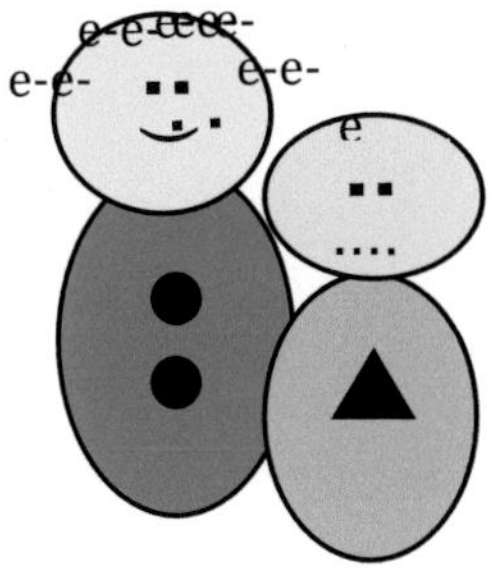

Neon approached the visitor standing in the shadows, questioning his presence on 8th Street. "Young man, what can I do for you?"

Sodium was almost speechless, overcome by the sharp contrast between 8th Street and the rest of Periodic Table Land. Finally, he managed to say, "I am fascinated by how happy you all are!"

Neon said, "Why should seeing us happy be so surprising to you? It's our nature to be happy. You act like you've never seen an atom that was happy before

Sodium said. "I haven't." He told Neon of the actual state of affairs throughout Periodic Table Land. "Absolutely not one atom outside of 8th Street is happy. My family lives on 1st Street, and we are all very sad. They sent me on a mission to find a happy atom and learn how to become happy. So I visited every family all across Periodic Table Land. All the way up to 7th Street the elements in the Chemical Families were sad."

"Now I come here and not one atom is sad. It is such a stark contrast. I'm sure you can understand why I am so amazed."

"Oh my!" said Neon, "I can't imagine anyone not happy. I guess when you are surrounded by all this happiness, it's hard to believe anyone could be sad."

Sodium asked if he might hang out and see if he could figure out why the Noble Gas family was so happy. He asked to stay and talk to all the elements on 8th Street. Neon made Sodium feel welcome. Leaving Sodium to do his investigation, Neon said again, "Oh my! How can any atom not be happy?"

Sodium moved through the crowd. At first he just walked around watching each of the elements in the Noble Gas Family. He greeted a few, and he tried to observe what they had in common. He believed logically, whatever they had in common was what made them happy. He decided to find some elements willing to talk with him. Finally he found two pleasant elements sitting on a bench where it was shady. Sodium went over to them and introduced himself.

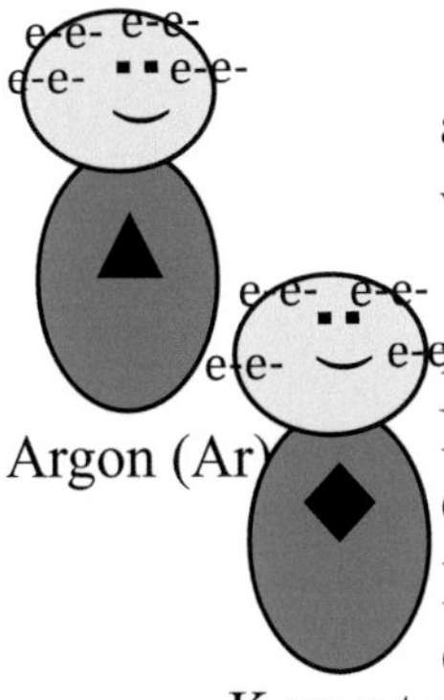

Argon and Krypton invited Sodium to sit down. He remained and talked with them for quite a while. They mostly told him what it was that they enjoyed doing. But this did not help.

After he finished visiting with Argon and Krypton, Sodium moved on. He continued to pursue the questions, "What do these Noble Gases have in common? What is it that makes these Noble Gases so different from all the other atoms in Periodic Table Land? If I can answer these two questions I will have the reason these Noble Gas elements are so happy."

Argon (Ar)

Krypton

Finally, Sodium came to the last two elements, Xenon and Radon. They were laughing, and talking with each other. Sodium approached them and had quite a long conversation with them. They were friendly, but they too did not know why they were happy.

Looking at them Sodium had an idea. It was more than and idea. Sodium was sure he had finally found the reason the Noble Gases were so happy. The idea came to him all of a sudden. All of the Noble Gasses have 8 electrons in their Outside Energy Level. That's why they're happy. None of the other atoms in all of Periodic Table Land have eight electrons in their Outside Energy Level. I've solved the problem. This explains why these Noble Gases are happy. They have 8 electrons in their Outside Energy Level." Sodium was overjoyed. "Hooray! I achieved my objective. I found atoms that were happy. Not only, did I find them, I figured out *why* they are happy."

Well, Sodium's theory was about to self destruct. In walks Helium whistling a happy tune, and smiling from ear to ear. He was as happy as all the other Noble Gases, but he had only two electrons in his Outside Energy Level. "How can he be happy if my theory is right? My theory must be wrong." Helium continued laughing and singing, and he had only 2 electrons. Helium ruined Sodium's theory that an atom needed 8 electrons in the Outside Energy Level to be happy.

Adding much to Sodium's confusion was the fact that the elements back on 2nd Street have 2 electrons in their Outside Energy Level, and they are extremely sad. "Why is this Helium atom so happy when he has only 2 electrons in his Outside Energy Level? Or should I be wondering why the atoms on 2nd Street, with 2 electrons in their Outside Energy Level, aren't happy at all? I know that the Alkaline Earth Metals on 2nd Street are sad. Here I believed I had solved the mystery of why the Noble Gases were so happy. Now Helium has ruined my theory. He's happy, and he has only 2 electrons in his Outside Energy Level. Oh my! My theory is not correct. I thought I could go home and tell my family that not only had I found a family that was happy, but also I learned why they were happy. Now I can't."

Along came Neon commenting, "Sodium, you looked so happy a few minutes ago, now you look sad again. What's wrong? I couldn't help noticing the confused and sad expression on your face. What's the trouble, my son?"

Sodium told him the whole story of how he thought he had figured out the reason the Noble Gases were happy. "I was sure it was because you all have 8 electrons in your Outside Energy Level, and none of the other elements in Periodic Table Land have 8.

Then Helium came in with only 2 electrons, and he was as happy as all the rest of the Noble Gases; but he did not have the 8 electrons that I believed an atom needed to be happy."

Neon said, "Don't worry. You were on the right track, but we are not happy **just because** we have 8 electrons in our Outside Energy Level. We are happy, because having eight electrons in our Outside Energy Level makes that level COMPLETE. Once the Outside Energy Level has 8 electrons in it, it can not hold any more electrons. It is completely satisfied. It's like the way you feel after a good meal, full, content and happy."

"Let me explain why Helium is happy with only 2 electrons."

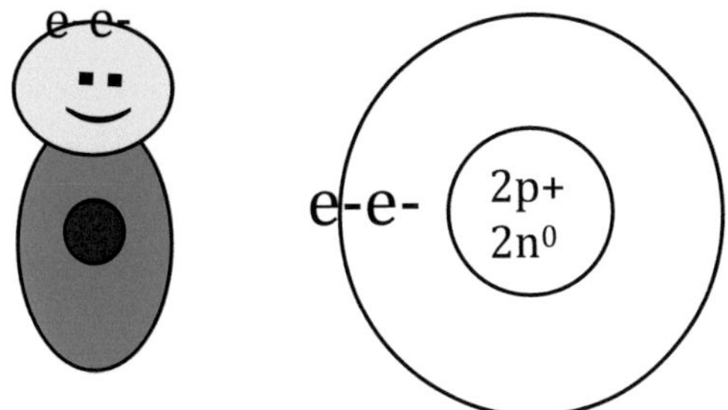

"Helium has only one energy level. Don't you remember from chemistry class that an atom with only one energy level is complete with only two electrons? Here in my chemistry book is a Bohr model of a Helium atom. Look at it. Helium has only one energy level, and you can see there are only 2 electrons in it. That energy level can not hold any more electrons. So, it is COMPLETE like all the other atoms that have 8 electrons in their Outside Energy Level. Helium's Outside Energy Level is full and complete with 2 electrons. So you did find the reason we Noble Gases are HappyAtoms."

Sodium said, "Thank you for showing me this. I really forgot. I was just not thinking of anything past the 8 electrons. I forgot that meant they had complete Outside Energy Levels."

"Mystery solved," said Neon. "Be happy Sodium. You found the secret of being a Happy Atom. **Atoms are Happy Atoms when they have complete Outside Energy Levels.** It's like feeling full and contented when you've had a good meal. You're satisfied. You figured out why the atoms here on 8th Street are such Happy Atoms. All our Outside Energy Levels are complete and totally satisfied."

This made Sodium feel so good. His mission was accomplished. He had found the secret of being a Happy Atom. The Outside Energy Level needs to be complete. **If an atom has only one Energy Level, it is complete with 2 electrons; if an atom has more energy levels, it is complete with 8 electrons.** Helium was happy, because his Outside Energy Level was complete with only 2 electrons in it.

Sodium thanked Neon for his help in learning why the Noble Gases were Happy Atoms. He couldn't wait to get home to tell his family that he had found Happy Atoms, lots of them, on 8th Street. Finding the secret of being happy could change the lives of everyone in Periodic Table Land. Sodium kept thinking, "It's simple. All you need to be happy is to have a complete OUTSIDE ENERGY LEVEL!" I found the Happy Atoms I was looking for. I even found out why they are happy."

All the little Noble Gases with their happy smiling faces gathered round to say goodbye to Sodium

Sodium took one last look at all those little Happy Atoms and waved goodbye. It was such a good experience to meet these Happy Atoms. Sodium said, "I will never forget how good it felt when I entered 8th Street for the first time—the laughter, the music, the happy talk." He walked back through the woods to find his magic bubble.Then began his long trip back to 1st Street. He couldn't wait to get home to share the good news with his family, The Alkali Metals.

All the way back, he kept remembering the smiling faces of the Noble Gas elements. He thought, soon the elements in my Alkali Metals will have smiling happy faces, too. This is my fond hope.

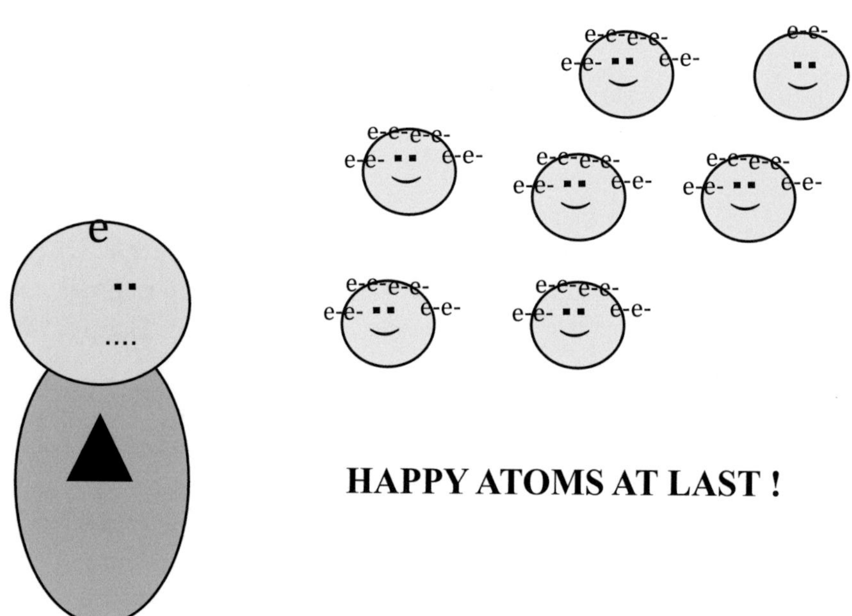

Sodium Returns Home With Good News
Chapter 4

Sodium's journey back to his family seemed to take forever. He viewed Periodic Table Land below with joy and hope in his heart. He had found the secret of becoming a Happy Atom. Now there was a chance for the Alkali Metals and all the families in Periodic Table Land to become happy. They would not have to be sad any longer. On his journey back Sodium made several important observations that he would remember later.

As he flew low over 7th Street with a perfect view of the Halogen family, Sodium could see that they had seven electrons in their Outside Energy Level. They just needed one more electron to have 8 electrons to be complete and happy like the Noble Gas family on 8th Street. A thought crossed Sodium's mind. "Each of the elements in my Alkali Metal family, has one electron in his Outside Energy Level. Maybe there would be a way for us to give the Halogens the one electron they need to become Happy Atoms." Sodium stored both these thoughts in his mind to think about later.

It seemed like the bubble was flying extra slow today, probably because he was in a hurry to tell his family the good news. Going slow did give Sodium time to think. As he passed over 6th Street, he noticed they had 6 electrons in their Outside Energy Level. Sodium thought, "If they could get just 2 more electrons, they would have the 8 electrons needed to have a complete Outside Energy Level and become Happy Atoms." As he passed by 5th Street, he noticed those elements had 5 electrons in their Outside Energy Level. His mind was processing thoughts rapidly. If they could get 3 more electrons, they would have the 8 electrons needed to complete their Outside Energy Level. Then they would become Happy Atoms, too. He stored all these thoughts in the back of his mind. They were facts that he might share later with elements that needed this information. Now he couldn't wait to get home to relate what he knew to his family. His mission was a huge success.

The whole Alkali Metal family was anxiously awaiting Sodium's return. Lithium, who had been the family lookout, saw the green balloons suddenly appear on the horizon. When he shouted out, "Balloons on the horizon!" Potassium, Cesium, Rubidium and all the other members of his family ran to the edge of 1st Street's landing strip. They were

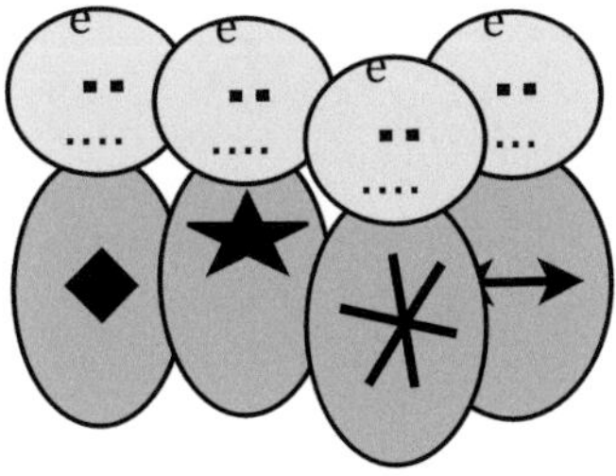

anxious to hear if Sodium had been successful in his search for a Happy Atom. Had Sodium found a Happy Atom? They were beside themselves with curiosity. They were anxious to escape the sadness that had trapped them for so long.

As Sodium climbed out of the bubble beneath the family balloons, they could tell by the expression on his face that something good had come of his investigation. Sodium yelled, “I did it! I found the secret of becoming a Happy Atom.” His words were tripping over themselves as they tumbled out of his mouth, attempting to recount the whole tale of his adventures in one breath. “Almost all the atoms in Periodic Table Land are sad. So, we are not the only sad atoms in the land. That’s what I found out first. I almost gave up hope. I thought I’d never find a Happy Atom. I had to go to the last street in Periodic Table Land before I ever found a Happy Atom. The Noble Gases on 8th Street were the only Happy Atoms in all of Periodic Table Land. I stayed there till I found out why they were happy. Are you ready to hear this? They’re happy because they have Complete Outside Energy Levels.”

Rubidium quickly asked, “What does that mean?”

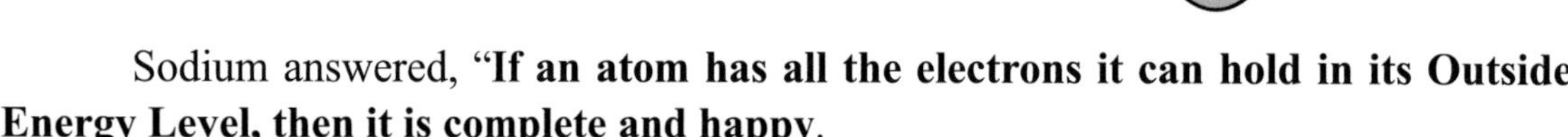

Sodium answered, “**If an atom has all the electrons it can hold in its Outside Energy Level, then it is complete and happy**.

An atom with only one energy level is complete when it has **2 electrons** there. When an atom is complete, it is happy. That’s what we want to be. Most elements have more than one energy level. So most atoms need 8 electrons in their Outside Energy Level to be complete and happy.”

Sodium directed, “Everyone check your Outside Energy Level and notice it has only one electron in it. We need seven more electrons to be complete. That’s a lot of electrons to get, and it’s hard to get that many.”

“On my return trip, I remembered what one wise old atom told me a long time ago. *Atoms try to do things the easiest way possible*. This made me think. Is there an easier way for the Alkali Metals to get a complete Outside Energy Level than to find seven more electrons? Then I decided to look at my own Alkali Metal atom to see if there was any other way. Look what I observed. I”ll draw a picture of my two last energy levels so everyone can see what I mean. Here it is.

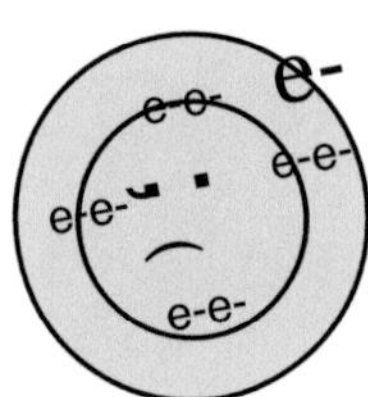

Radium said, “I understand. We have a complete energy level beneath our present Outside Energy Level which has only one electron in it. If we give that electron away, our hidden complete energy level will pop up. Then we’ll have a complete Outside Energy Level. If it works, we will become Happy Atoms because we will have a complete Outside Energy Level. What a great plan!”

Potassium questioned, "Do I have it right? Our Outside Energy Level has only one electron in it now. If we give away that electron, we will be left with a complete Outside Energy Level. It will be complete with 8 electrons. That means we will be Happy Atoms. Is this what we will look like? And will we have a happy face?"

"Right on Potassium! You got the idea. If we learn how to give away our outside electron, we will all become Happy Atoms with a big happy smile. Make sure each of you understands that if we give away that one electron, presently in our Outside Energy Level, our hidden complete energy level will pop up and become our Outside Energy Level. Then, we will have a complete Outside Energy Level, and be happy like those Noble Gases on 8th Street."

"Even you, Lithium, will be happy if you give away that outside electron. You will be left with only one energy level. When you have only one energy level it is complete with only two electrons. Here's what you will look like."

"Are you sure Sodium? I'm really worried."

Sodium tried to assure Lithium. "I was on 8th Street when Helium walked in with only 2 electrons, like the 2 you have. Because he had only one energy level, he was as happy as everyone else. The reason he was happy was because elements with only one energy level are complete with just 2 electrons. So I'm positive, you will be happy, Lithium. You will be smiling like everyone else. Absolutely!"

Now, Sodium continued saying, "On my way home I observed something that gave me an idea. I noticed that the *Halogens* had 7 electrons in their Outside Energy Level. They need to have 8 electrons there to be happy. That means, they are *missing one electron* to have a complete Outside Energy Level. We have that one electron that the Halogens need. If we give the Halogens our 1 electron, they will have eight electrons in their Outside Energy Level. They will be complete and happy. When we give away our one electron, our complete energy level will pop up. Then we will have a complete Outside Energy Level, and we'll be happy too."

"Here's my plan. I will go to 7th Street and talk to Chlorine. He's the Halogen I know best. I will explain to him the secret of becoming happy. When he learns that he needs 8 electrons in his Outside Energy Level to be happy, he will see that he needs one more electron. I'll show him the one electron that I'm willing to give him, and we'll see what happens when he takes my one electron. If we both become Happy Atoms, I'll come back, and get you. Then, we can all go to 7th Street and become Happy Atoms with all the Halogens. I really feel this is going to work."

All the Alkali Metals cheered. "Three cheers for Sodium! You're the best."

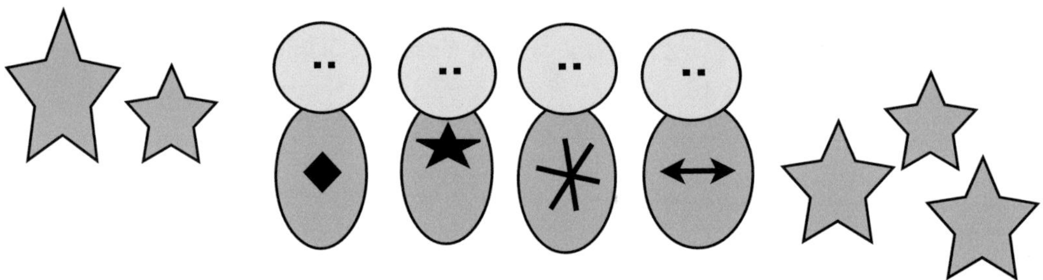

Sodium said, "Don't cheer for me yet. This is only a **hypothesis, a guess about what logically should happen.** The next step every scientist must take is to test the hypothesis. **Testing the hypothesis is called experimentation**. My plan explains how the experiment will work. You can cheer for me if Chlorine and I become Happy Atoms."

****

Sodium knew he had to share the good news with Professor Terry. After all, she was the one who motivated him to find a way for his family to be happy. Sodium headed to the chemistry lab at the university. Professor Terry was sitting at her desk doing some calculations related to her latest experiment. Sodium surprised her with his visit. "Both you and Guy need to hear the good news. I followed your suggestions and found the secret of being happy! All any element needs to be happy is to have a complete Outside Energy Level." He went on to relate in detail his adventures visiting all the families in Periodic Table Land and finally finding Happy Atoms on 8th Street.

Sodium said, "I am ready to take this one step further. I have a plan to actually *become* a Happy Atom. I'm about to leave for 7th Street, the home of the Halogens to test my hypothesis. Let me explain my plan as simply as possible.

"The elements in my Alkali Metals family have one electron to give away. If we give away that one electron, our hidden complete energy level will pop up and become our complete Outside Energy Level. This will make us happy. The Halogens have 7 electrons in their Outside Energy Level. Since they need 8 electrons to be happy, taking our one electron will make them happy, too. Since they are tired of being so sad, I'm sure the Halogens will agree to try this simple experiment."

"I'm going to visit Chlorine first, and I would like to invite you and Guy to be there to witness this exciting moment when Chlorine and I could possibly become Happy Atoms. Meet me over in Periodic Table Land at my home, #11, 1st Street as soon as possible. Would first thing tomorrow be possible? I'll be waiting for you both. We can all go together to visit Chlorine and you can watch us try my experiment."

Professor Terry thanked Sodium for the invitation saying, I'm glad you thought to invite us. Tomorrow morning should work. It is important for Guy to observe you becoming a Happy Atom. By becoming a Happy Atom, you will form a compound, and compound formation is what I planned to teach Guy next. Seeing you exchange electrons with Chlorine will be the perfect way for Guy to learn how compounds are formed."

Sodium left and returned home questioning the successful outcome of his proposed experiment——"Will Chlorine take my one electron? Will we both become Happy Atoms? I sure hope so. Actually trying out this great idea will give us the answer to that question. I hope my experiment will be successful. I found a happy atom; I found our why atoms were happy.Next is what all my effort was about—becoming a Happy Atom."

Professor Terry sent a text to Guy telling him to get ready to return once again to Periodic Table Land early in the morning. They would be learning how sad little atoms become Happy Atoms. He said, "If Sodium's experiment works, we will be right up close where we can watch how a compound is formed. This is what you are well prepared to learn next."

****

When it was dark enough to see the stars, Guy opened the back door and walked up the mountain trail behind his parent's cabin careful not to trip over the roots and ruts in the path. The smell of pine made him happy. He finally found the rock that was so perfect to lean back on to view the night sky. He gazed at the constellations and was amazed as always at the vastness of the universe. He thought about the Bohr models of

the atoms which are the same out there in space as they are here in these rocks and trees. As Guy watched the shooting stars, the tree beside him began to shake. Guy looked up, saw the branches of the tree spread apart, and said, "Is that you Wish Star?" Down came his dear friend Wish Star to sit beside him.

Wish Star said, "I'm so excited for you Guy. I heard you're going to Periodic Table Land to see Sodium and Chlorine finally become a Happy Atoms. That will be such a great experience. I hope you know that when Sodium and Chlorine become happy, a compound is going to be formed. Now you're getting into real chemistry. I am happy that you are ready to understand compounds. I'm proud that you've learned so much. Just remember, I'm always here for you."

Wish Star sat with Guy a long time looking at the beautiful night sky. Then he tapped Guy with his magic wand wishing him success in his new adventure.The visit over, Wish Star flew off into the night sky.

As Guy watched, Wish Star turned and waved to him one last time. Guy sat there staring into that majestic sky imagining the planets whirling around in space. He thought of the Bohr models he drew, the Chemical Families he met and was just spell bound thinking of how the entire universe was made up of only 92 of the 118 elements on the Periodic Table. Now I'm going to learn how these elements join together and become compounds. It will be so good to watch sad little atoms become Happy Atoms. Like the ones on 8th Street I can't wait.

****

Meanwhile, Sodium was back in Periodic Table Land making plans.He kept thinking about those Noble Gas elements who were so happy. He couldn't wait to visit the Halogens and see if his family really could become Happy Atoms.

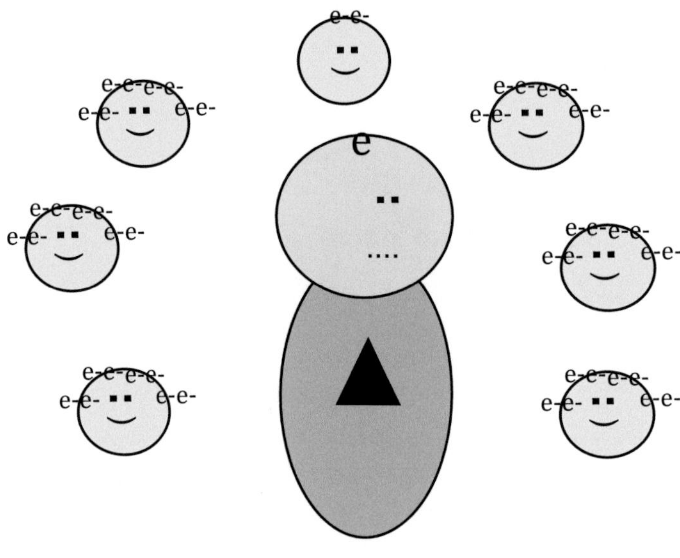

Wish Star says, "You've read **BOOK 2.** You drew the **Structure of Atoms**. You met the **Chemical Families.** You learned the importance of the electrons in their **Outside Energy Level**.

Now, it's time to read **BOOK 3**. Learn how each of the **C**hemical **F**amilies form **Compounds.** Find out how **F**orming **C**ompounds changes sad little atoms into **Happy Atoms**.

Learn to **W**rite and **U**nderstand

Chemical Formulas.

Continue to learn Basic Chemistry.

The Happy Atom Story is

MAGICAL !

Printed in the United States
By Bookmasters